Dr. Ashley M. Berge, PhD
AM8 International
QLD, AUSTRALIA
+61 401 814 324
AM8@topicthread.com

AM8International.com

WHAT IS YOUR GAME MISSING?

The Scientific Truth in Breaking Down Tennis' Elite

By
Dr. Ashley M. Berge, PhD

First published in 2019 by Dr. Ashley M. Berge, PhD
https://AM8International.com

2nd Edition
ISBN: 978-0-9945694-7-9

DEDICATION

To Malibu,

And the players who yearn for the top, the coaches who want to get you there, and the parents who want you to navigate the injuries and reach the pinnacle of the game.

There is only one good, knowledge, and one evil, ignorance.

– Socrates

FOREWORD

The attainment of knowledge is a selfish feat and one left with moral repercussions. Whether to keep to oneself and die whole with this knowledge inside; or, to disperse this knowledge and face the insurmountable ineptitude of those arguing that science is not synonymous with factual truth. One is an arguably selfish cause for withholding what others can gain from, whilst the other is a path filled with repercussion and turmoil for opting not to withhold what has the capacity to enrich the lives of millions around the world. If you're reading this then you're well aware the later has been the path chosen – a selfish choosing in itself for the cliché making *the world a better place* despite the throw-back.

Science is a complicated art that often is misunderstood more often than not as a result of over-complication. The art of simplification of science is an equal if not more enduring art that permits a wider reach to harness the science at hand. As with previous texts, **What is Your Game Missing?** is no different from embedding this art to allow the dispersion of knowledge to be accessible to all – but that's where the similarities stop.

What is Your Game Missing? was intended to be kept private – withheld from the world as a consequence of the already tireless repercussions of hundreds too naïve to accept the *truth* of science. They didn't deserve to know. But, this would come with its own set of consequences. Posed with the question: *what about the non-ignorant or non-naïve who accept science for its factual value – the millions around the world opposed to the mere hundreds?* Confronted with a moral choice, statistics were called on – science was the drawing board. Numerically, the naïve represented less than 1% of the cumulative total. Statistically, the naïve are outweighed by the remaining 99% – concealing the *truth* on account of the minority *from* the majority no longer held reason.

Originally the concept of The V by Dr. B © was born close to two decades ago – to be discussed inside. It took near that amount of time for the scientific voyage to be undertaken and for the *truth* to be unveiled. **What is Your Game Missing?** is an ultimate correlation from its birth to its unveiling (Part 1 and Part 3) and to its real-world application (Part 2). Despite the years between the unveiling and its continued application, the preservation of science is amplified. Meanwhile, as the world changes around us there is one unifying factor that remains and waits for it to be built upon – *science* literally makes the world go 'round.

A simplified recount of this enduring voyage is partially provided for pretext:

How I came to be, *recap*:

> At 18, I became a Personal Trainer / Sports & Strength Conditioning Coach. I was also a full-time athlete, training for 8 hours a day.
>
> At 19, I was a sponsored Elite Athlete. I also commenced University and maintained my full-time training.
>
> At 20, I became a Level 1 Coach, then went on to become one of Australia's youngest ever Level 2 Coaches permitted to go back-to-back.
>
> Just before I turned 21, I received a scholarship to play and train at one of America's leading Academies, in Texas.
>
> At 21, I was coaching some of America's top national players as a direct result of succumbing to another injury and whilst also playing and training when healthy. That same year I received a full-scholarship to a National Collegiate Athletics Association (NCAA) DII private University that marked against the Harvard Grading Scale.
>
> At 22, I spent part of the summer touring through Europe with American national players. At this time I had also started my Undergraduate degree and had received a Grade Point Average (GPA) of a 4.0 (the highest attainable) in

both first and second semesters. This came with multiple Dean's awards.

At 23, I again spent part of the summer touring through Europe with American national players. Due to injury, I had to fly back to Australia and leave my scholarship behind. I spent the majority of the year on crutches unable to walk. This was the year I completed my Undergraduate degree and started to run. This was also the year I had to choose between playing and coaching — I chose coaching.

At 24, I commenced my Masters. I spent the summer, again, touring through Europe with American national players. On return I was asked to be the Director of a tennis program and accepted, being one of the youngest coaches, nationally, to take on such a position. After running for less than a year I managed to clock up 50 km days (not weeks). Just before I turned 25, I managed my biggest week yet: 180 km's in total.

At 25, I registered my first business with my Dad and opted not to travel but to keep on top of my Masters (after receiving a GPA of 7.0: the highest attainable). Shortly after, my Dad passed away — just shy of my graduation; four months later, my Grandpa passed away.

At 26, I competed in my first Ultra race (100 km) but had a DNF (Did Not Finish) due to an ongoing hamstring injury. In the same year, I became recognised as a Level 3 Coach (the highest attainable globally). Soon after, I commenced my PhD — I still had questions to answer that my studies until then were unable to address.

At 27, my PhD was under full-swing – in not one, but between *two* Schools (a result of the depth of the work undertaken) and I was on track to complete my PhD in the minimum time frame. The same year I was able to sponsor my first athlete.

A fortnight after turning 28, I became unwell. Ultimately, I couldn't run, walk, read, write, do simple tasks, let alone finish the home-stretch of my PhD without near-collapsing. Without going into detail, somehow I did manage to finish my PhD and in the minimum time frame; with only four months leave taken. For an elaboration, *The Secrets to Optimal Wellbeing* shares more.

At 29, I had PhD corrections to look after, although the entire year was all about getting my health back. That same year, I competed in my first Ultra since becoming unwell — and I finished all 80 km's of it, even though I tore my tendon at the 20-25 km mark. It was about finishing and I did. Mid-year, I had surgery that could have left me far-worse off (from my ongoing health battle) and I am incredibly fortunate to still be here.

At 30, I had already published the first instalment in *The Secrets to Optimal Success* series: *The Secrets to Optimal Performance Success*, and later that year, *The Secrets to Optimal Wellbeing* was published. And, almost two years since submission, I was officially awarded my doctorate. It was also a year of tightening the vision for what the science inside my PhD was, and eventually over time, going to bring the world. As a result, *The Science of Elite Performance* was published just shy of turning 31.

At 31 it was all about bringing this vision into fruition – and so the journey began of building a Social Compilation, Topicthread, that brought the world together, using the *two* disciplines of my doctorate, Health *and* Education, with *performance* its overarching tone, as the vehicle to unify individuals across the world. And with that, 'social' was at its core – using the scientific model inside my PhD. This saw Topicthread (topicthread.com) come to life.

At 32, *The Secrets to Optimal Coaching Success* was published, and slowly but surely, Topicthread was the World's 1st Social Compilation – and the 1st interface with the architecture to deliver access to Health *and* Education, in unison, universally. Meanwhile, quietly in the background I had been working on *The Elite Research Method*, which was subsequently published later that year.

And now, on the cusp of turning 33, **What is Your Game Missing?** lays before you, and the 6th publication with AM8 International. Most significantly, **What is Your Game Missing?** is the capstone of the unveiling of the scientific undertones that came forward through my research and its landmark findings in elite sporting performances, better known as The V by Dr. B ©.

The V by Dr. B © is an original work of science – one day to be built upon. Like the thousands of scientific findings prevailing around the globe, they all started somewhere – whether as an original or an *add-on* to its respective origin, likened perhaps to a *donut with its now varied glazes*. A consequence of choosing to disperse the knowledge of The V by Dr. B © is great hesitation. At times like this, I am reminded that knowledge comes with power, tied to an unwavering responsibility – to enlighten, to share, to evolve, and again, whether naïve or not, to *make the world a better place* for the millions who want to know, and deserve to know, the *truth*.

An underlying reason behind the choice in my early twenties was the consequence of succumbing to a number of injuries that saw myself sidelined each and every time I managed to climb back to my peak – and the want to stop others from enduring a similar feat. The V by Dr. B © is identified as a functional movement parameter, for the first time disclosed – a movement that uses the body's natural fluidity to achieve an ultimate performance outcome, **without** *compromising the body and contributing to a player succumbing to injury.*

The functionality of The V by Dr. B © thus is a movement and performance parameter that works with a player's body and **drastically** *limits the onset of injury when applied – it's that simple. My hope and wish for The V by Dr. B © is to find its way into the hand of the millions of aspiring coaches and players alike to not only aide in optimal performance outcomes, but equally provide longevity of performance whilst reducing the onset of injuries. I hope this* **missing** *piece of your* **game** *brings you to new heights, whether as a player or coach, and this new scientific* **truth** *sees you embrace The V by Dr. B © now and always through play.*

CONTENTS

List of Figures

WHAT IS YOUR GAME MISSING?

The Scientific Truth in Breaking Down Tennis' Elite

By
Dr. Ashley M. Berge, PhD

INTRODUCTION

Tennis has inevitably *forever* changed. At the hands of science and *scientific discovery*, tennis was the sport of choice to dismantle – it just so happened through this process *landmark* implications for the sport were unveiled. Tennis will never be the same. The *historical* implications of the scientific discovery set the foundations for the game to come – unrivalled by any scientific evidence before it, **inside** places in print for the *first* time the *truth* behind elite performances.

Historically, tennis is a game of finesse and anticipation. The fluidity of performance *was* in a class of its own. Players learned to use their body in sync with their racket to achieve the desired result. Although rigid, players were too in a class of their own as the fundamentals of movement were being discovered. Born was the role of biomechanics and how it feeds into functional movement parameters, or patterns.

Change was inevitable. There was more to be desired. Speed of play increased. The power behind the ball became an ingenious way of climbing to the peak. A combination of speed and power was unrivalled. At the same time, players were at the mercy of their body. A player's body was at the mercy of their biomechanics and their functionality. The onslaught of injuries to hit the game saw players succumb to the fate of misused technical arrangements. The optimal technique eluded the game – more apparent than ever as a need to be harnessed. If anything, for the fate of the players and their longevity in the sport.

Populated over the years was the forehand groundstroke. Whilst other aspects of a player's game were highlighted, the forehand remained as the most dominantly used. The volley was initially a favoured weapon, yet as speed and power were increasingly introduced, the volley was no longer most dominant. A transition was ignited between the one-hander which then became two – no longer were single-handed backhands as common; players struggled to keep speed and power without the addition of the added hand. Meanwhile, a consequence of the added hand noted a restriction in the ability to produce **angular play** – heightened functionality of movement was paramount.

The serve was the original weapon. It still is, for some. However, the anticipation of play is more crucial than ever with speed and power as the deciding denominator in a player's results. The returner now has a greater chance of success if their groundstrokes are in tune with their anticipation. Meanwhile, one universal factor remains – the forehand groundstroke is off a player's dominant hand and is more significantly applied single-handedly; in contrast, the backhand groundstroke has its variations in single and double-handed use – the forehand trumps in uniformity of application.

A direct implication of uniformity and functionality values the forehand groundstroke in large over the backhand groundstroke. Fundamentally, the forehand groundstroke affords transference, whereby the backhand has its limitations. **The Rule of Transference** is hereby introduced: *a functional application of one movement and its technical parameters are inevitably transferred as a consequence of conditioning, whereby a player is conditioned to perform a set task accomplished through the sound execution of the conditioned technique and its sets discrete skills. Thereby, these skills are inevitably tied to the players rudimental skills of technical performance parameters and their body's functionality to perform them. Thus, The Rule of Transference is deeply rooted in the players conditioning of a skill and inevitably its technique. With likened properties, this skill is thus transferred in its application and through its very conditioning, aiding a player's ability to perform a set or* *given technique against its predecessor – contributing in large to the output of optimal performance parameters.*

Tennis in the *twenty-first* century has changed as a result of human adaptation. Speed and power now go hand in hand. The game is quicker than ever before. The role of anticipation is crucial. A player's body is placed under strenuous pressures no other generation is yet to endure. Interestingly, no other generation endured the mercies of the game at the heights of the present generation. Injuries run rife. Frequently a wide scale of players endure through a season whilst managing an injury – set to bring their season to its knees. The advancement of the game and the role of speed and power hasn't been without its victims.

The V by Dr. B © previously referenced in earlier texts, is introduced in full for the *first* time inside – for the flailing player it may be a saviour. For the generation to come, The V by Dr. B © is a necessity for survival at the elite ranks and how to climb to the top. Furthermore, **The Rule of Transference** affords ease of technical conditioning and is profound in the rudimentary uptake of The V by Dr. B ©. Whilst in conjunction to the *Optimal Performance Theory* introduced in *The Science of Elite Performance* (Berge, 2017), **What is Your Game Missing?** cements the basis of functional movement whilst applying **The Rule of Transference**.

The origins of tennis are no different than other sports – players learned not only by concept, but how their bodies naturally prevailed through performance. As insinuated, players had an aspect of flow – rigid nonetheless. As the generations adapted, so did the game – inevitably, speed and power were a must. This brings us to *now* and their prevalence. Likewise, it brings us to *now* and the scale of injury. No longer is a band-aid fix acceptable. A functional movement that aligns with and upholds **The Rule of Transference** is a necessity for progress. And thus The V by

Dr. B © brings about the next chapter of change – inevitable to the game, for progress to rein.

In order to feed a thriving generation of optimal performance parameters, The V by Dr. B © is introduced and its singular and mixed criteria. **What is Your Game Missing?** is presented in three distinct forms. Initially, *The Unveiling* is shared – a first in print to be disclosed in order to provide not only the truth behind elite performance, but the scientific discovery that is directly responsible for the historical implications for tennis as it is known. But, application is key. Too often scientific discovery is scarce from real-world application and how it is currently established. Behind *The Unveiling* then comes *Coaching Analyses* for the application of The V by Dr. B © in the real-world and its use – or failure (of-use) in player performance.

Sandwiched between *The Unveiling* and *The Statistics* is the present *truth* of the game's current elite players – on the WTA (Women's Tennis Association) and ATP (Association of Tennis Professionals) tours, for female and male players, respectively. *The Statistics* is the final piece inside **What is Your Game Missing?** that affords transparency in numbers – tied to *The Unveiling*. Intentionally placed apart, *Coaching Analyses* brings **The Rule of Transference** and its scientific discovery denoted by The V by Dr. B © to life, whilst *The Unveiling* and *The Statistics* solidify the implications of science, respectively, with the later disclosing additional data sets.

Change is finicky, largely a consequence of the *new* explained by human psychology as the often discomfort associated with newness, whether scientifically discovered, or not. **What is Your Game Missing?** pre-empts this through its sandwiched approach in the unveiling of The V by Dr. B © and its descriptive statistics – often *not* what the human psyche responds towards; rather, its application in the real-world – thus, lastly shared. Understanding of concept and scientific discovery is translucent with ease of uptake – thus, initially shared. Recall, change is inevitable. Change affords progress. Meanwhile, scientific discovery expands our universal knowledge, delivering answers never once known – to the readily consumed. For the *first* time and now open for the taking, The V by Dr. B © is unveiled and the race to elitedom is in your hands.

PART 1

THE UNVEILING

Part 1: Introduction

The sporting world doesn't know what's coming. Nor do they know what is now in your very palm. Often a select few are privy to the truth and do one of two things. One, they do nothing – often a consequence of the greatest minds falling on deaf ears: the unbelievers of *truth*. Two, they abuse it. The former is a moral high ground against the later, often a consequence of the *truth* kept from sight. It's a game in itself.

Awareness of the game affords you a step ahead to play it on your terms. *The Unveiling* is the fine line that exists *between* One and Two, an unparalleled *truth* and historical first in elite sports performance. Specifically, the unveiling of a core proponent with a direct correlation – statistically significant in itself, on what distinguishes the best players in the world from the rest – and how it can simply be integrated into a player's **game**, now, as a **missing** piece to their optimal performance success.

The foundations need to become apparent. In its simplest sense, prior to this voyage The V by Dr. B © was merely a theory – yet to be proven. In order to prove that something exists, science is relied upon rigorously, stretched beyond comprehension, meticulously dismantled before being placed back together, and only then, to see if the shoe fits – metaphorically speaking. If it doesn't – better luck next time. If it does, you have cause to dig deeper. Fundamentally, mere cause in itself isn't enough. Raw data is demanded and that data better be statistically significant, or else – the theory remains an ideological concept without a scientific explanation to move on. The data always trumps; in this case, no fewer than 121,770 statistical inferences.

Recall, science is factual evidence; recollect, science is *truth*. No one sets the rules themselves – the rules are set by science.

If it doesn't adhere, that's final. And *specificity* is key – tennis, the sport; the forehand groundstroke, the technique. Articulated in its simplest sense, the forehand groundstroke relies on a players most dominant hand and is transferrable in its application. Technically, the backhand varies in its discrete skills per player preference and its non-transference as it *crosses* the player's body. The forehands discrete skills *are* transferable to the backhand and its *sister-ball-sports*. A magnitude of *specificity* is the selection of the forehand for its transference and its now solidification in the theory behind the *unveiling* of The V by Dr. B ©.

Disclosed in *full* transparency is the statistically significant data that brought The V by Dr. B © to life – alongside its counterparts. All scientific findings have their respective origins and *The Unveiling's* disclosure is no different. Contextually, the data individually showcases the discrete skills that comprise the serial skill aptly named – the *fore*hand groundstroke. Scientifically, literature is called on for introductory terms of reference, albeit scarcely, to feed back into the contextualized unveiling. Fundamentally, the capstone that continues to segment the elite is unveiled. With aforementioned hesitation, The V by Dr. B © begins.

Unearthing The V by Dr. B ©

The unearthing of The V by Dr. B © was dependent on three specific questions in alignment with its unique performance indicators, and parameters respectively:

1. To what extent is knee bend prevalent at point of contact in the forehand groundstroke on hard court surfaces in elite tennis players?
2. At what height is the ball at point of contact in the forehand groundstroke typically contacted on hard court surfaces in elite tennis players (in relation to hip height — a visual reference point)?
3. What is the correlation between prevalence of knee bend at point of contact on hard court surfaces in elite players, and increased performance (higher ranking), which would indicate if The V by Dr. B © is prevalent in elite players?

Whilst the forehand groundstroke formed the foundations for the performance parameters, it in itself was the *vehicle* at large – transferrable to the backhand groundstroke, as with its sister-sports that apply its associated technical parameters.

A historical unveiling for sports performances around the world, The V by Dr. B © is at the very essence of the correlation between an athlete's body, their ultimate technique through performance, and their relationship with an object i.e. a ball. Totalling in excess of 121,000 individual notions, The V by Dr. B © captures landmarks findings in sport and **What is Your Game Missing?** is the very first window into what the best players in the world do differently, consistently – securing their status as one of the best players in the world – referred to as an elite *elites* i.e. the elite *of* the elite.

The data were generated using notational analysis, a form of video and in-person analysis, discussed by Hughes and Franks (2004) as an objective data collection tool in analysing components of an athlete's technique. Video analysis enabled the observation and notation of tennis players' performance based on a customised observation tool. Notational analysis also allowed identification of movement patterns against a set of criteria in real-time observational analysis, that is, from a live court-side perspective. The use of live and video-based observations allowed consideration of external factors that may have impeded the live observational process and exclude any conflicting data (see Part 3 for *The Statistics*). As such, the inclusion of both forms of observation would increase the trustworthiness of the observational tool (Gratton & Jones, 2010).

The first question, introduced, was concerned with both the existence and frequency of The V by Dr. B © as applied by elite tennis players during competition, when contacting their forehand groundstroke. Three criteria were analysed using notational observation: knee bend, balance, and ball height at point of contact. All balls were required to land in the singles court to count and be marked against the criteria. As a result, three criteria were charted against a unique observation chart and the resulting frequencies recorded. The separate frequencies of these three criteria are reported first in *The Unveiling* as the first step in identifying the existence of The V by Dr. B © in the forehand groundstroke of world-class players.

To further identify The V by Dr. B © in action, the mixed criteria were analysed. The mixed criteria included combinations of knee bend and point of contact to illustrate their applied frequency among the world's elite players. The mixed criteria were representative of The V by Dr. B ©: the consideration of both knee bend and height of point of contact during the contact phase of the forehand groundstroke. Its relevance is embedded in the very nature of the introduced questions, and by considering more than one variable at a time allowed The V by Dr. B © to be considered as a whole (with both criteria) compared to singularly recording these respective variables.

To find ranking commonalities in the application of The V by Dr. B ©, the results of the criteria (singularly and mixed) were subsequently correlated against player rankings to identify if there is a significant relationship between these two variables. Player rankings with the single and mixed variables were analysed, followed by a representation of the elite, specifically, the top 10 players on both the WTA (women's) and ATP (men's) tours. These rankings and the frequency of knee bend and ball height, at point of contact in the forehand groundstroke, were then reported. *The Unveiling* concludes with exploring if there are gender differences between male and female players when executing The V by Dr. B © during forehand groundstroke play, or rather, application.

The following sections detail the process of notational analysis and the scientific parameters. The reliability of the notational analysis in terms of inter-rater reliability findings will be reported before the prevalence of The V by Dr. B © will be identified in elite tennis, and its application during the forehand groundstroke.

The Process of Notational Analysis

The observational tool was designed to record three criteria (or performance indicators), knee bend, point of contact and balance. The three criteria can vary in terms of the applied variable. These variables are: deep knee bend, slight knee bend, no knee bend; contact above hip height, contact at/around hip height, contact below hip height; and, balance or not balanced. The available sample included 400 players in total, with a selection range of 1 to 200 on either the WTA or ATP tours. Tournament entry conditions limited the sample size to <250 at the Australian Open, <100 at the Brisbane International, and <20 at the Hopman Cup (2012—2019). The results of *The Unveiling* include a sample size of 123 participants, carefully analysed against the set criteria. This predetermined sample was a consequence of tournament schedules for live analysis and commercial television broadcasts of matches. Thus the sample was selected with these constraints in mind (Long, 2007). The performance analysis criteria (or parameters) included rules required by each participant as disclosed in *The Elite Research Method* for inclusion.

Thirty forehands were selected to ensure a consistent number of forehands were recorded. In some cases, matches were found to go from start to finish with less than 30 forehand groundstrokes hit. Such participants were excluded from the overall results. Although these matches were still observed as there was no way to know if a match/player would fall under the 30 benchmark. Ultimately, in excess of 100 player observation charts were excluded due to the total number of shots falling short of the required 30 forehand groundstrokes observed. This meant that the 30-ball margin was high with 100 plus players observed falling below 30 forehand groundstrokes hit within the duration of a match.

The findings included only those who frequently hit their forehand groundstroke (at least 30 times in a match), minimising the likelihood of recording an unfavourable stroke. Overall, the forehand groundstrokes recorded against the criteria totalled 11,070 notations across three separate criteria: knee bend, balance, and height of point of contact (in addition to shot outcome), totalling 121,770 cumulatively.

To complement the unveiling of the historical data and its performance-specific implications, a further 99 players were analysed in *Coaching Analyses* against the criteria through a compilation of the notational analysis in-action – forming a specific performance analysis accompanying the data set. The *performance specific analysis* aligns with The V by Dr. B © featuring players on both the ATP and WTA tours largely holding a ranking at time of analysis inside the top 100 on either tour, respectively.

Reliability of the notational analysis

In order to establish and report on inter-rater reliability, a group of select matches ($n = 5$) were notated by two independent raters: the scientist and an experienced research assistant sought to complete the observation tool. The two raters observed the identical match/individual at the same time but independently. Collectively, the results of the combined matches ($n = 5$) were used to evaluate reliability of the observation tool. This required both data sets to be analysed against the other. The data set included total notations for the criteria of balance ($n = 120$) for the variables of balanced and not balanced observed over the 30 instances by both researchers. Totals for the criteria of knee bend ($n = 180$), including the variables of no knee bend, slight knee bend, and deep knee bend were then reported over the 30 instances observed by both researchers. Totals were then identified for the criteria of height of point of contact ($n = 180$) and the variables contact below hip height, contact at/around hip height, and contact above hip height were then reported. The totals for each player ($n = 5$) were analysed using the Kappa statistic through SPSS. Viera and Garrett (2005) report the Kappa value of 0.81-0.99 to have an almost perfect agreement, values between 0.61-0.80 as a substantial agreement, 0.41-0.60 to have a moderate agreement, whilst lower Kappa values (i.e., <0.41) are considered fair through to less than a chance of agreement. These values will be used to evaluate the Kappa values of the present unveiling.

The average Kappa reported for the knee bend criterion was 0.86, representing strong agreement between both raters in using the observational tool to rate the variables of knee bend. Overall, both researchers reported the use of knee bend the same 82% of the time. Next, the average Kappa found for the balance criterion was 1.0, representing perfect agreement between both raters for these criteria. The third criterion, height of point of contact, was found to have an average

Kappa of 0.69, which represents a moderate agreement between the two raters. However, when considering the reporting of height of point of contact individually, both scientist and research assistant were in agreement 79% of the time.

It is evident from the Kappa results that balance is the criterion with the best agreement between raters, followed by knee bend, then height of point of contact. This could be explained as a result of the speed of play, and how the inclusion criteria and boundaries were understood.

Notational Observational Findings

The observation results were gathered through a strict data entry process recorded against three distinct independent variables (knee bend, height of point of contact, and balance) to account for one result (with the fourth variable of ‘in’ mandatory for the shot to be notated) tabulated for each player on 30 individual occasions throughout a single top tier tennis match. The independent variables knee bend and height of point of contact had three selections; whereas balance had two (see Figure 1). Balance was either recorded as "balanced" or "unbalanced" in contrast to knee bend, comprising no knee bend, slight knee bend or deep knee bend, and height of point of contact as above hip height, at/around hip height, or below hip height. For a visual reference, Figure 1 consists of images produced through Kinovea to best illustrate the application of knee bend and the height of point of contact in the forehand groundstroke whilst the player maintains (dynamic) balance. The results generated from the notational analysis follows, beginning with the prevalence of knee bend.

Figure 1 Grigor Dimitrov, a currently active player (left) and Agnieszka Radwanska, a recently retired player (right) demonstrating a balanced state due to the presence of knee bend (slight and deep, respectively), with a height of point of contact below hip height

Source: © Network 10 Australia and/or Seven Network Australia

The V by Dr. B © — A Statistical Representation of Prevalence

The V by Dr. B © is a *scientifically discovered* lower body position at point of contact that occurs during the action required to contact the forehand groundstroke. The V by Dr. B © is a position that comprises of three variables: 1) identifiable knee bend, 2) height of point of contact around or below a player's COG, and 3) dynamic balance — a stable base of support, for instance, the player's position is not compromised. The V by Dr. B © is identifiable through the application of knee bend *and* height of point of contact that falls within the range of at or below hip height, creating a V shape in three areas at the same time (Figure 2 and Figure 3):

1. At point of contact, the angle generated by hip flexion resulting in the slope of the quadriceps (primarily at the leading leg) and the player making contact with the ball (to the side) at/around or below hip height, creating a V shape,

2. At contact, the angle generated behind the leading leg (closest to the racket hand) and respective knee joint creates a V shape,

3. At contact, the angle created between the leading leg (closest to the racket hand) respective ankle and shin.

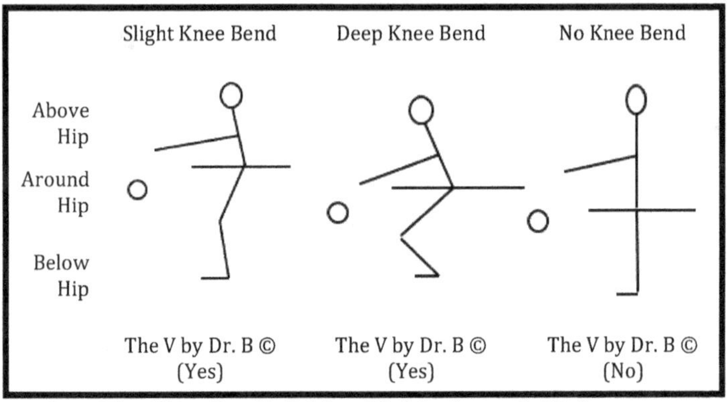

Figure 2 The V by Dr. B © identifiable through V shapes

Figure 3 What does The V by Dr. B © look like? Simplified illustration of The V by Dr. B © (top), followed by Roger Federer, a currently active player (bottom) captured with a height of point of contact below and slight knee bend

Source: © Seven Network Australia

The V by Dr. B © assists tennis players to perform the optimal forehand groundstroke. Ultimately, the statistics disclosed will provide clarity on best-practice of how to apply The V by Dr. B © in the forehand groundstroke, currently not included in the literature. Throughout the observation process, criteria of The V by Dr. B © were identified using notational analysis, recording against a unique Observation Chart. The data for each variable were entered into SPSS to generate descriptive statistics (including frequencies) and compute correlations (as required). The following sections report the individual statistical analyses for each variable (singularly and as mixed criteria) from the notational analysis (totalling 121,770) of the 123 players, beginning with knee bend.

Knee Bend

One of the defining criteria of The V by Dr. B © is knee bend at point of contact during the execution of the forehand groundstroke. The two variables that represent the application of the knee bend criterion, are slight and deep knee bend. According to Schonborn (2000), ideal knee bend is around 120 degrees. From this work, slight knee bend was defined and denoted by the angle behind the leading leg, corresponding with the racquet arm (e.g. the right leg for a right-handed player), as being >121 degrees. The cut-off angle of slight knee bend is exhibited by the presence of no knee bend. Whilst the third variable, deep knee bend, is denoted by the angle behind the leading leg, corresponding with the racquet arm, being <120 degrees, with the cut-off angle beginning at 121 degrees or greater. The following figures illustrate when players exhibit slight knee bend and deep knee bend in the forehand groundstroke. The findings are representative of the 30 groundstrokes recorded for each 123 players and the identified level of knee bend when players made contact with the ball in the forehand groundstroke. Figures 4, 5 and 6 were generated using Kinovea software to best represent and provide a visual reference of the criterion: knee bend, and its associated three variables.

Figure 4 Andy Murray, a currently active player with slight knee bend

Source: © Seven Network Australia

Figure 5 Agnieszka Radwanska, a recently retired player with deep knee bend

Source: © Network 10 Australia and/or Seven Network Australia

Figure 6 Grigor Dimitrov, a currently active player with deep knee bend

Source: © Network 10 Australia and/or Seven Network Australia

No Knee Bend

When generating frequency statistics from the Observation Chart to report the occurrence of no knee bend at point of contact in the forehand groundstroke, *no* evidence of no knee bend was found. That is, all the observations revealed that players *always* bend their knees when contacting the forehand groundstroke (see Part 3 for further Statistics).

Of strong significance, no knee bend was *not* observed throughout the entire observation process, due to the alternative of knee bend being present. As knee bend *was* observed in the 123 players, the characteristic of The V by Dr. B © is a consistent variable present in the forehand groundstroke in professional tennis players. The next section will consider the frequency of slight knee bend in the forehand groundstroke.

Slight Knee Bend

The variable of slight knee bend was first looked at to try to explain what made up the remainder of applied knee bend, given that no knee bend was not recorded. It was of interest to find the type of knee bend most commonly applied. Frequency statistics were generated for the occurrence of slight knee bend during the contact phase of the forehand groundstroke and found the mean to be 7.3 ($SD = 0.5$) out of the 30 balls observed throughout the observation process (see Part 3 for Complete Statistics). Figure 7 illustrates the histogram with frequencies 5 and 7 as the most commonly occurring numbers of slight knee bend.

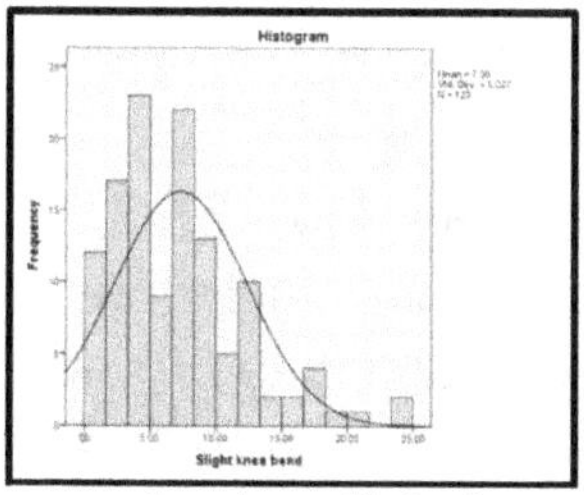

Figure 7 Histogram showing slight knee bend in the 123 players observed

Figure 7 identifies slight knee bend at point of contact, illustrating that players are bending their knees when contacting their forehand groundstroke on average 7.3 times out of a possible 30 balls observed. The result indicates that the only alternative lower body position is deep knee bend due to zero evidence of *no knee bend* being recorded throughout the observational period, explaining the remaining 22.7 shots. The next section examines deep knee bend and its frequency.

Deep Knee Bend

When looking at deep knee bend, its frequency was of concern due to the 'no' and 'slight' variables not having a substantial mean. However, it was found that the frequency statistics for the occurrence of deep knee bend during the contact phase of the forehand groundstroke had a mean of 22.7 ($SD = 4.9$) out of all participants analysed (see Part 3 for Complete Statistics). The histogram with frequencies in the range of 22–27 as the most common occurring are illustrated in Figure 8.

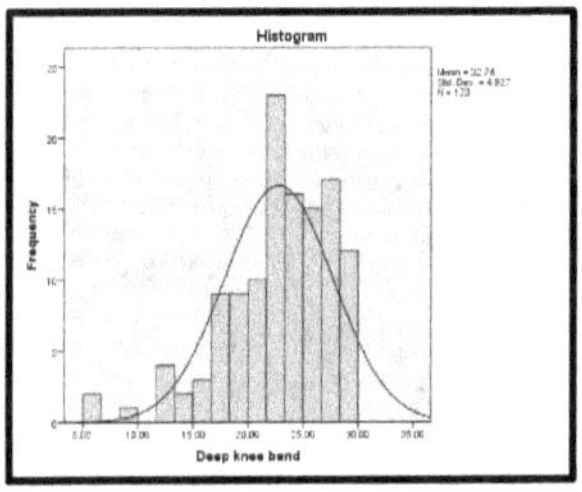

Figure 8 Histogram showing deep knee bend in the 123 players observed

Through the analysis of deep knee bend, players bend their knees (deeply) when contacting their forehand groundstroke on average 22.7 times (range = 5 to 30) out of a possible 30 balls observed. The frequency of deep knee bend increases as the total ascends to 30, whereas two participants were noted as demonstrating deep knee bend 12 times out of a possible 30, seven participants applied deep knee bend 23 times (out of a possible 30), and nine participants applied deep knee bend 27 times (out of a possible 30). This indicated that *deep knee bend* is a consistent movement applied by players when contacting the forehand groundstroke.

The analysis of knee bend and more specifically the strong evidence of deep knee bend being the ideal lower body position during the forehand groundstroke is important to the development of an optimal point of contact position. The use of knee bend supports the argument that as players bend their knees deeply at point of contact in the forehand groundstroke, that in order to make contact with the ball, the *height* of the ball at point of contact is controlled by the player. This is a result of the player's footwork, allowing them to modify their positioning to coincide with the adopted level of knee bend, without compromising shot outcome.

The height of point of contact is determined by the player and their state of balance, ensuring the player is able to perform the required technique of the forehand groundstroke. Having identified that deep knee bend is the dominant position during the forehand groundstroke and that, in turn, has implications for the height of point of contact. This suggests that this height will also reduce (be lowered) as a result of the player lowering their centre of gravity (COG), a result of deep knee bend promoting a balanced and thus stable position. The criterion of balance and how dynamic balance contributes to the overall performance of the forehand groundstroke and plays a crucial role in the application of deep knee bend and height of point of contact will now be reported.

Balance

The results thus far indicate that the forehand groundstrokes observed occurred with either slight or deep knee bend, resulting in a lower body position. It can be asserted that elite players are lowering their COG as they make contact with the ball. As established by Garhammer (1989), lowering one's COG increases stability which can also be observed as being balanced. This section is concerned with notational analysis around player balance, with the variables of balanced or unbalanced noted for these criteria. The criteria devised for balanced or unbalanced, included a simple separation. If the player's position was observed to be compromised, for instance, they were out of position and the player was unbalanced. If the player was either falling forwards, sideways or backwards due to unstable footwork, this was also classed as unbalanced. A player was noted as balanced if they were able to cleanly execute their forehand groundstroke without their position being compromised, for example, they were ready to hit the next ball if necessary and maintained court position.

Denoted by The V by Dr. B ©, balance is a characteristic dependent on players exhibiting dynamic balance at point of contact during the execution of the forehand groundstroke. The statistics presented in this section summarise the incidence of balance or unbalance in the forehand groundstroke. The results have been generated on these criteria, representative of the 30 balls recorded for players observed at point of contact in the forehand groundstroke. Figure 9 depicts an image generated to best represent and provide a visual reference of this criteria, with all images shown also illustrating the continued presence of either deep or slight knee bend.

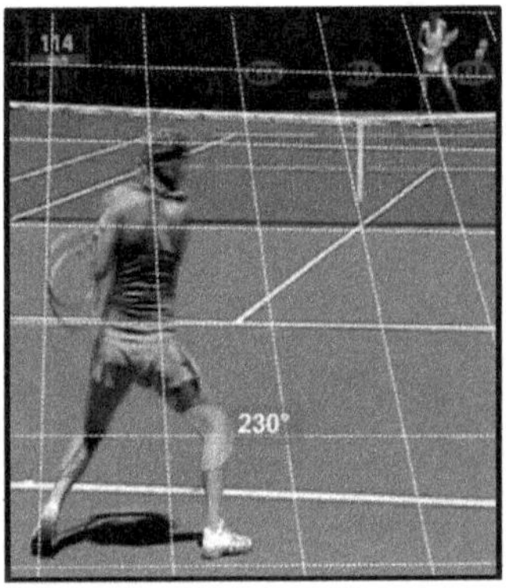

Figure 9 Victoria Azarenka, a currently active player illustrating dynamic balance when making contact with their forehand groundstroke

Source: © Seven Network Australia

Balanced

In order to differentiate between dynamic balance and an unbalanced position during the forehand groundstroke, frequency statistics were generated to illustrate the presence of balance amongst elite tennis players at point of contact (Figure 10). The mean was found to be 28.8 ($SD = 1.310$) out of the 123 participants observed (see Part 3 for further Statistics).

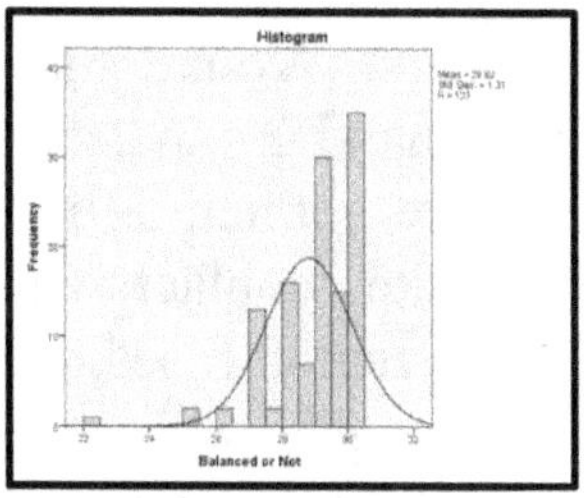

Figure 10 Histogram showing balance in the 123 players observed

Players were found to exhibit a balanced state during contact with the forehand groundstroke on average 28.8 times out of a possible 30 balls observed (range = 22 to 30). The results serve to indicate that knee bend and balance occur together, and discussed by Garhammer (1989), knee bend results in a lowered COG, found to be consistent in elite players when contacting their forehand groundstroke. The presence of dynamic balance not only complements the role of knee bend, but knee bend is dependent on the presence of dynamic balance, as is the height of point of contact in the forehand groundstroke. Height of point of contact and how the third criteria is reliant on the results of *Knee Bend* and *Balance* will now be discussed.

Point of Contact

The V by Dr. B © is possible only when players exhibit a point of contact at/around or below the hip region, otherwise termed in the vicinity of their COG, during the execution of the forehand groundstroke. Height of point of contact falling around or below a player's COG conforms with the characteristics of The V by Dr. B ©. Height of point of contact, the third criteria of The V by Dr. B ©, is of concern to identify if its characteristics align with already discussed results of *knee bend* and *balance.*

The figures presented in sections *Contact above hip height, Contact at/around Hip Height* and *Contact below hip height* illustrate the variables of height of point of contact respectively (above hip height, around the hip region, and below hip height) in the forehand groundstroke. The height of point of contact variables are defined as follows: above hip height, when the player makes contact with their forehand groundstroke above their hip region (used as a visual point of reference); at/around hip height, when the player makes contact with their forehand groundstroke around hip height, also referred to as COG (used as a visual point of reference); or below hip height, when the player makes contact with their forehand groundstroke below their hips (used as a visual point of reference). The results are representative of the 30 balls recorded for each player observed making contact with the ball at point of contact in the forehand groundstroke. Figure 11 depicts an image generated using Kinovea software to best represent and provide a visual reference of the variable.

Figure 11 Li Na, a former elite elites and now retired captured with a height of point of contact below hip height

Source: © Seven Network Australia

Contact above

When looking for height of point of contact in the forehand groundstroke to be identified above hip height, its frequency was of concern, not exhibiting characteristics of The V by Dr. B ©. In order to identify height of point of contact in the forehand groundstroke for this variable, frequency statistics were generated of contact above hip height during the contact phase of the forehand groundstroke. The mean was found to be 3.0 (*SD* = 2.2) out of all participants observed (see Part 3 for further Statistics). The histogram with frequencies 1, 3 and 4 as the most commonly occurring is illustrated in Figure 12.

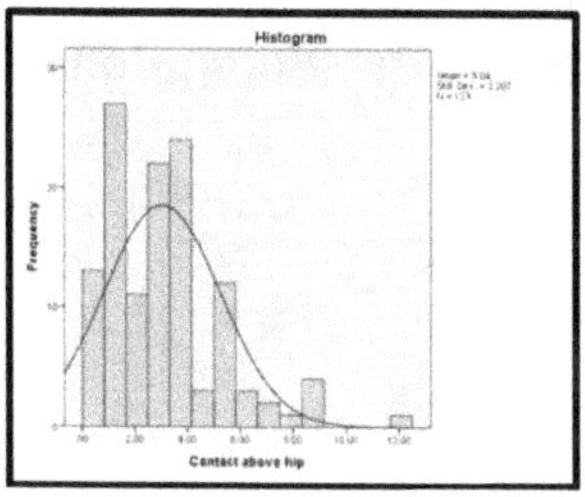

Figure 12 Histogram showing contact above in the 123 players observed

Height of point of contact above hip height was found to occur on average 3.0 times (range = 0 to 21) out of a possible 30 balls observed. This indicates that on average 10% of shots occur above hip height and the remaining 90% can be suggested as conforming to characteristics of The V by Dr. B ©. The strong evidence of height of point of contact being subsequently (90% alternative) either at/around or below a player's COG during the forehand groundstroke is important to the *scientific discovery* of an ideal position at this phase of the stroke sequence – The V by Dr. B ©. The results suggest that this height of point of contact, limited in its application, supports the argument that as players bend their knees considerably (as found in *Knee Bend*) at point of contact, in order to make contact with the ball, the height of the ball at point of contact is adjusted by the player's movement to ensure it falls within their hitting range. The alternative variables will now be discussed that exhibit traits of The V by Dr. B ©, beginning with height of contact at/around a player's hip height.

Contact at/around Hip Height

To identify the frequency of point of contact to be located around hip height in the forehand groundstroke, frequency statistics were generated to identify if this height of point of contact was applied by the elite participant group. The mean was found to be 7.3 (*SD* = 3.1) out of all participants analysed (see Part 3 for further Statistics). The histogram with frequencies 6, 7 and 8 as the most common occurring is given in Figure 13.

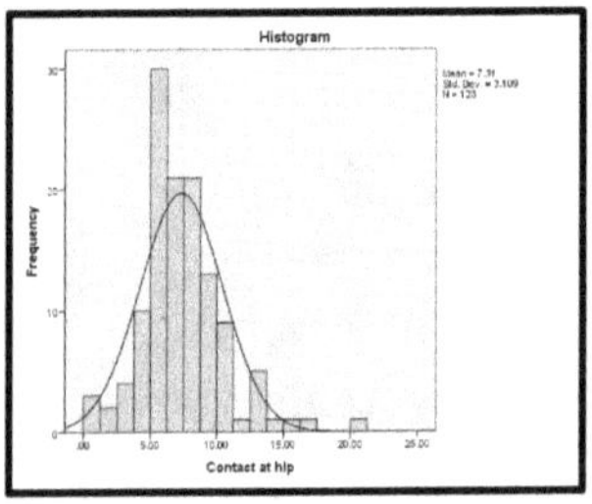

Figure 13 Histogram showing contact at/ around hip in the 123 players observed

The finding of height of point of contact occurring around hip height on average 7.3 times (range = 0 to 21) out of a possible 30 balls observed, aligns with characteristics of The V by Dr. B ©. The results of this variable, although occurring less than 25% of the time, suggest the remaining statistics will also align with The V by Dr. B © (accounting for 65% of results). This result is substantial, contributing towards an ideal position to be applied at the point of contact phase in the forehand groundstroke technique, *now* scientifically known as The V by Dr. B ©. The results indicate a lower height of point of contact is most optimal among the world's best tennis players, in contrast with the data displayed in Contact Above Hip Height. The results of contact below hip height and its consequent alignment with The V by Dr. B © will now be discussed.

Contact below hip height

Looking to build on the results of *Contact above hip height* and *Contact at/around hip height*, accounting for the remainder 65% of results at point of contact in the forehand groundstroke, frequency statistics were generated to determine the frequency of contact below hip height during the contact phase of the forehand groundstroke. The mean was found to be 19.6 ($SD = 4.0$, range = 6 to 30) out of all participants analysed (see Part 3 for further Statistics). The histogram illustrates frequencies 18, 22 and 23 as the most common instances in Figure 14.

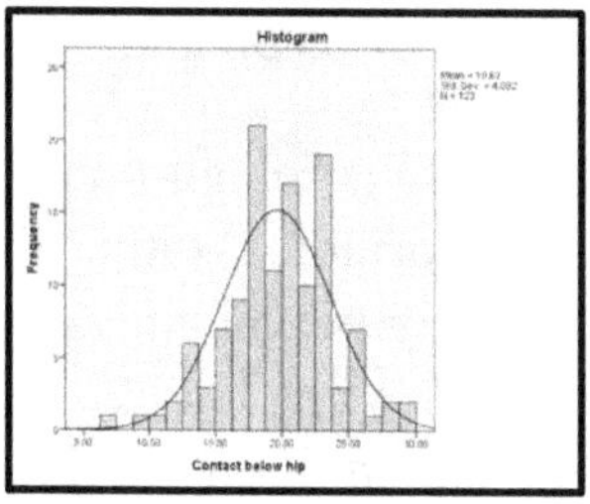

Figure 14 Histogram showing contact below hip in the 123 players observed

Height of point of contact in the forehand groundstroke is found to be most frequent at below hip height: occurring on average 19.6 times out of a possible 30 balls observed. This result indicates that elite players make contact with their forehand groundstroke below hip height on average 65% of the time, a much higher frequency than contact above hip height or contact at/around the hip region. The strong evidence of height of point of contact being predominantly below a player's COG during the forehand groundstroke is significant to the advancement of an ideal position, *known* as The V by Dr. B ©. Through Point of Contact it was found that the results suggest a height of point of contact below a player's COG supports the argument that as players apply deep knee bend (as found in Deep Knee Bend) at point of contact, the *height* of the ball at point of contact is lowered, and consequently, contacted most consistently below a player's hip region: illustrating the characteristics of The V by Dr. B ©.

The descriptive statistics thus far have detailed the three criteria and their respective variables of The V by Dr. B ©. The presented statistics contribute to the *scientific discovery* of The V by Dr. B © by noting the incidence and frequencies of its variables. Although the variables of The V by Dr. B © that are in fact most frequently applied by the world's elite tennis players have been found singularly, the results in combination (mixed); that is, height of point of contact *with* knee bend and the frequency of results, are yet to be shown.

Ultimately, The V by Dr. B © is reliant on the combination of the three criteria and their variables, rather than individually. Therefore, although the most prevalent variable for each of the three criteria associated with The V by Dr. B © have already been defined, and *scientifically discovered*, the results serve to indicate the presence of each variable alone, not their combination or in fact if a relationship exists. It is now the intention to look at the combinations, expressed as the mixed criteria of these variables, and to identify if the most frequented variables reported singularly, deep knee bend and height of point of contact below hip height, are as frequent in combination when players make contact with the forehand groundstroke. The mixed criteria (Mixed Criteria: Knee Bend (deep/slight) and Point of Contact (above/hip/below)) will now be discussed in terms of the variables that exhibit traits of The V by Dr. B ©.

Mixed Criteria — Knee Bend (deep/slight) and Point of Contact (above/hip/below)

The V by Dr. B © is applied when players exhibit evidence of knee bend, balance *and* make contact at/around or below hip height when contacting the forehand groundstroke. The figures presented in this section will illustrate, compare and contrast incidence of these variables against one another, for instance slight knee bend and deep knee bend *with* point of contact at/around hip height and below hip height, respectively, at point of contact in the forehand groundstroke. The criterion of balance has been excluded however due to a lack of variability in results across all players observed. Figures 15, 16 and 17 display images generated using Kinovea software to best represent and provide a visual reference of the mixed criteria and the combinations of these variables. The findings are representative of the 30 groundstrokes observed for each of 123 players, and the classification of the variables when making contact with the ball. The results on slight knee bend and contact around hip height (Slight Knee Bend/Contact at/around Hip Height) will first be presented.

Figure 15 Li Na, a elite elites at time of the initial investigation, now retired, showing slight knee bend and a height of point of contact below hip height

Source: © Seven Network Australia

Figure 16 Roger Federer, a current elite elites showing slight knee bend and a height of point of contact below hip height

Source: © Seven Network Australia

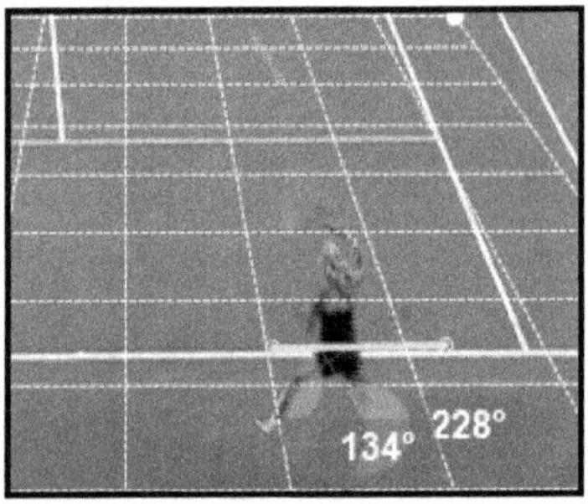

Figure 17 Dominika Cibulkova, a currently active player showing deep knee bend and a height of point of contact around hip height

Source: © Network 10 Australia and/or Seven Network Australia

Slight Knee Bend/Contact at/around Hip Height

When generating frequency statistics for the occurrence of slight knee bend and contact at hip height during the contact phase of the forehand groundstroke, it was of concern to identify if the results reflected those of the variables when they were considered singularly. The mean came to 2.6 (*SD* = 2.4) out of the 123 participants analysed (see Part 3 for further Statistics). The histogram with the frequency of 3 as the most common occurring number (zero excluded) as depicted in Figure 18.

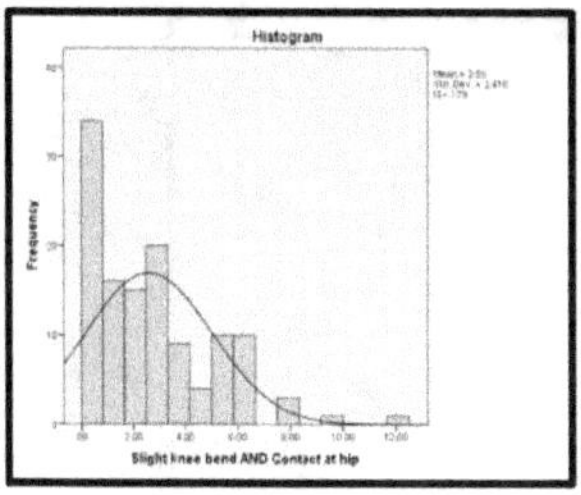

Figure 18 Histogram showing slight knee bend and contact at/around hip in the 123 players observed

The identification of slight knee bend coinciding with point of contact around hip height illustrates that players are using this combination when contacting their forehand groundstroke on average 2.6 times (range = 0 to 12) out of a possible 30 balls observed. Players were found to apply this mixed criteria on average less than 10% of the time when hitting their forehand groundstroke. The identification of this combination is of significance as it indicates that characteristics of The V by Dr. B © are not only applied, but at the same time, in unison, when elite tennis players make contact with their forehand groundstroke. Although this mixed criterion is not very high in occurrence, the results suggest that the other 90% of the time elite players are applying greater depths of The V by Dr. B ©, for instance a deeper knee flexion and a lowered height of point of contact. The results for the mixed criteria of slight knee bend and height of point of contact below hip height will now be shown.

Slight Knee Bend/ Contact below Hip Height

When looking at the frequency of slight knee bend and its association with height of point of contact below hip height, it was of concern to see if this combination was more frequent given that the variable below was the most frequent variable for height of point of contact singularly. Frequency statistics were generated for the occurrence of the mixed criteria, where the mean was found to be 2.0 (*SD* = 3.3), illustrated by the histogram in Figure 19 (see Part 3 for further Statistics).

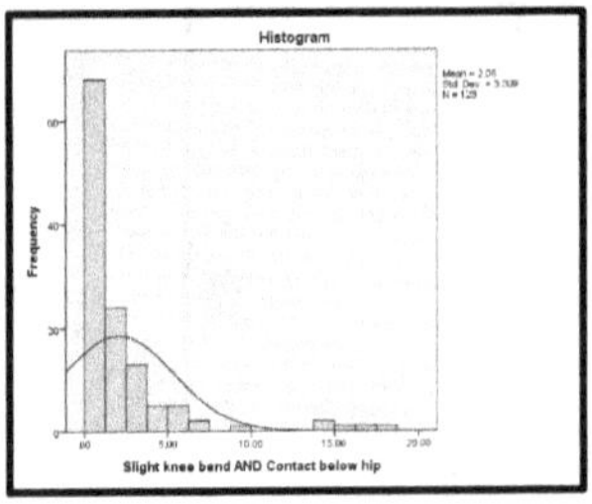

Figure 19 Histogram of slight knee bend and contact below hip in the 123 players observed

The mixed criteria occur on average 2.1 times (range = 0 to 19) out of a possible 30 balls observed. This mixed criterion, applied less than 10% of the time, exhibits characteristics of The V by Dr. B ©, though it is less frequent than the results of the variables individually. The results suggest that the more frequent variable: deep knee bend, opposed to slight knee bend, is associated with optimal performance, accounting for the height of point of contact to be lowered. It is noted that this is why this mixed criteria is not frequently applied due to deep knee bend not being present. Deep knee bend and its relationship with a height of point of contact at hip height will now be discussed in order to identify the most frequently applied variables of the mixed criteria present in the forehand groundstroke.

Deep Knee Bend/ Contact at/around Hip Height

When considering the frequency of deep knee bend in relation to height of point of contact at hip height, it was of concern if deep knee bend impacted the overall result in contrast to slight knee bend. Frequency statistics were generated for the mixed criteria and the mean was found to be 4.7 (*SD* = 3.1), illustrated by the histogram in Figure 20 (see Part 3 for further Statistics).

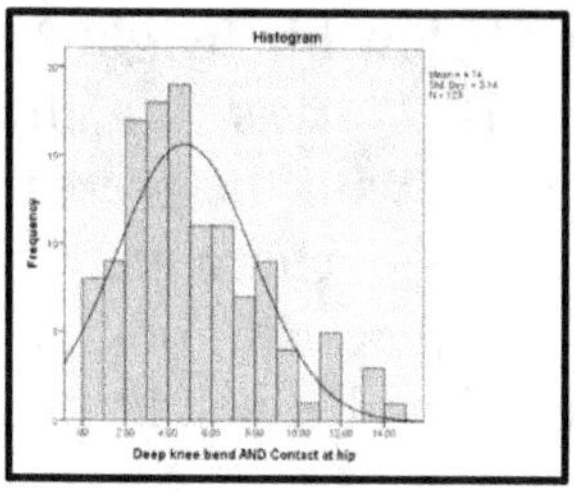

Figure 20 Histogram of deep knee bend and contact at/around hip height in the 123 players observed

The results of the mixed criteria indicate that the presence of deep knee bend increases in frequency, in contrast to the application of slight knee bend (Slight Knee Bend/Contact at/around Hip Height) when considering the variable of height of point of contact at/around hip height. The mean of deep knee bend is 4.7 (range = 0 to 15) in contrast to the mean of slight knee bend at 2.6. The result indicates that players apply deep knee bend *and* contact around hip height less than 16% of the time when hitting their forehand groundstroke. Currently the mixed criterion stands at 8%, 7% and 16% respectively when variables were combined together, increasing to 16% when deep knee bend is considered. When a height of point of contact below hip height (the most applied variable) was associated with slight knee bend, the results were not as favourable. The results suggest that the final mixed criteria, considering both variables found to be applied most frequently singularly, deep knee bend and height of contact below hip height, will account for the most frequently applied mixed criteria (excluding height of point of contact above hip height, due to its exclusion when considering the characteristics of The V by Dr. B ©). The results of deep knee bend and a height of point of contact below hip height (Deep Knee Bend/Contact Below Hip Height) will now be presented and its application when players contact their forehand groundstroke.

Deep knee bend/contact below hip height

When The V by Dr. B © was initially a *theory* the characteristics were to align with the presence of knee bend and a height of point of contact at/around or below a player's COG. It was therefore of interest to look at this association, the use of deep knee bend and a height of point of contact below hip height.

To identify the use of this mixed criteria, frequency statistics were generated where the mean was found to be 17.6 ($SD = 4.6$) out of the 123 participants analysed (see Part 3 for further Statistics). The histogram, displaying a bell curve with the frequency of 17 as the most common occurring, is depicted in Figure 21.

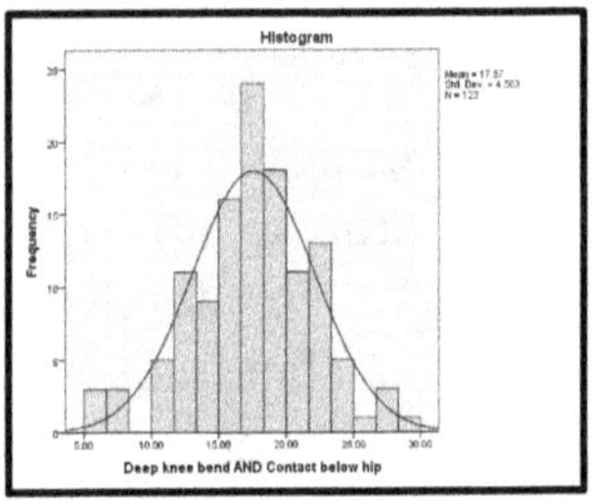

Figure 21 Histogram showing deep knee bend and contact below hip height in the 123 players observed

As displayed in the histogram, the results illustrate that when the mixed criteria of deep knee bend and contact below hip height are combined, it is the most frequently applied position performed by the elite players when contacting the forehand groundstroke. The identification of deep knee bend coinciding *with* a height of point of contact below hip height illustrates that players are using this combination when contacting their forehand groundstroke on average 17.5 times (range = 5 to 30) out of a possible 30 balls observed. The frequency of this mixed criterion was found to account for 59% of all mixed criteria when considering the most frequently applied position at point of contact in the forehand groundstroke. This result is substantial, outweighing the other alternatives of the mixed criteria, 8%, 7% and 16% respectively (the remaining 10% represents the use of either slight or deep knee bend when considering a height of point of contact above hip height: excluded due to it not aligning with the characteristics of The V by Dr. B ©).

When considering the mixed criteria, the results align with the earlier results of the variables singularly (Deep Knee Bend and Point of Contact Below Hip Height, respectively). When the most frequently occurring variables are considered in combination, the results are important (occurring 59% of the time). Both variables, deep knee bend *and* height of point of contact below hip height, align with the components of The V by Dr. B ©. The overall statistics indicate that when deep knee bend is considered with either below or at/around heights of point of contact, the frequency of use increases. This is opposed to that of slight knee bend and its association with the height of point of contact variables. This reveals that when players lower their COG, they are also lowering their height of point of contact.

Having identified the presence of The V by Dr. B © among the world's elite when contacting the forehand groundstroke, attention is brought to player performance and the correlation between the three criteria and player rankings. By looking at player rankings and the frequency of The V by Dr. B ©, it is then established if the frequency of the combined variables correlates with player rankings, indicating a statistically significant relationship between The V by Dr. B © and success in the forehand groundstroke.

Player Rankings

When The V by Dr. B © was theorised, it was concerned with the frequency of its characteristics among the world's elite players. This elite status would ultimately illustrate a successful position that is being used by the world's best when executing the forehand groundstroke. Although it has been found through Deep Knee Bend that the frequency of deep knee bend, and contact below hip height, the frequency of height of point of contact to predominantly fall below a player's hip height, as the most dominantly applied variables, and through Deep Knee Bend/Contact Below Hip Height as the most dominant applied mixed criteria, the association of The V by Dr. B © with player rankings is yet to be examined. Through this section (Ranking) the characteristics of The V by Dr. B © will be addressed, singularly and as a mixed criterion, and if a correlation is present with player rankings. The findings are representative of the 30 balls observed for 123 players when making contact with the forehand groundstroke.

Ranking

All players ranked inside the top 200 on either the WTA or ATP tours were included in the notational analysis. The observation process was indiscriminate to ensure a wide range of players were analysed and not a particular group favoured. When generating frequency statistics for player rankings in the participants observed during the contact phase of the forehand groundstroke, it was of concern to ensure a range of players was represented. The results show the ranking mean to be 70.7 ($SD = 57.8$) out of the participants analysed (see Part 3 for further Statistics). The histogram illustrates the top of the curve to be approximately 70, representing the mean in player rankings observed and the participants ranked inside the set-margin in the observation process are depicted in Figure 22.

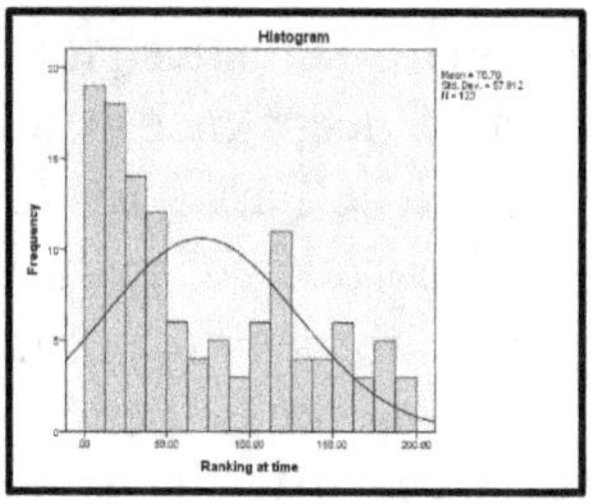

Figure 22 Histogram showing player rankings in the 123 players observed

With the average player ranking observed to be 70.7 (range = 1 to 200), a wide range of players have been represented, aligning with the set criteria for inclusion (ranking range 1–200). This indicates that the participant pool represents the assertion that elite players were included in the notational analysis. Additionally, it was identified that players inside the top 100 have been more frequently observed than those outside this ranking range, ensuring the focus is on the world's best tennis players. The results indicate that players who fall outside this range (101–200) have also been considered to ensure that a reasonable comparison in rankings and performance can be identified. The relationship between player rankings and parameters of the most optimal variables of The V by Dr. B © and its characteristics will now be discussed. For all other parameters, and criteria of The V by Dr. B © and their relationship with player rankings see Part 3 for further Statistics.

Ranking and Deep Knee Bend

The variable of deep knee bend has been shown to be the most frequently applied position when analysed singularly. It has also been found, thus far, that deep knee bend is also the most favourable position when considering the mixed criteria. Player rankings are yet to be correlated with deep knee bend. It is of significance to identify if deep knee bend is in fact a position used more frequently by the elite echelon of the playing group.

When generating descriptive statistics for player ranking (70.7) and the application of deep knee bend, the mean was found to be 22.7 ($SD = 4.9$) (see Part 3 for further Statistics). Player rankings were then correlated against the number of 'deep knee bend' in the forehand groundstroke, where Pearson's $r = 0.178$ for deep knee bend with $p = 0.048$ statistical significance (2-tailed). The result shows a significant but weak correlation between ranking and the number of deep knee bend in the forehand groundstroke. Presented below, in Figure 23, is the scatter plot providing an illustration of the results.

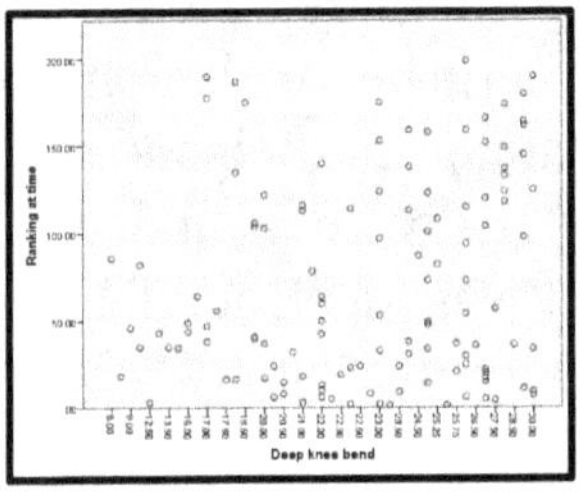

Figure 23 Scatter plot of ranking at time and deep knee bend in the 123 players observed

Through the results, it has been identified that deep knee bend does in fact account for the remaining knee bend applied by the elite (22.7 times out of a possible 30 balls), indicating that knee bend is a consistent element in a player's technique when contacting the forehand groundstroke. Significantly, the application of deep knee bend by the world's best players (those closer to number one), appears to be more prevalent than those with a lower ranking (closer to 200). The results of deep knee bend align with the results in Deep Knee Bend and the presence of knee bend in the elite technique of the forehand groundstroke – aligning with the application of The V by Dr. B ©. Height of point of contact below hip height and its correlation with player rankings will now be discussed.

Ranking and Contact below Hip Height

When considering a height of point of contact below hip height, it was of concern to identify not only if the results aligned with those in Point of Contact, but also if the results accounted for the currently unaccounted (19.7) outcomes out of the 30 observed groundstrokes. Descriptive statistics were generated for player ranking (70.7) and point of contact below hip height and the mean was found to be 19.6 ($SD = 4.0$). Player rankings were then correlated against the number of 'point of contact below hip height'. Pearson's correlation $r = -0.321$ for contact below hip height with $p = 0.000$ statistical significance (2-tailed) (see Part 3 for further Statistics). The result is a statistically significant negative relationship in player rankings, and players making contact at a height of point of contact below hip height in the forehand groundstroke (see Figure 24 for the scatter plot).

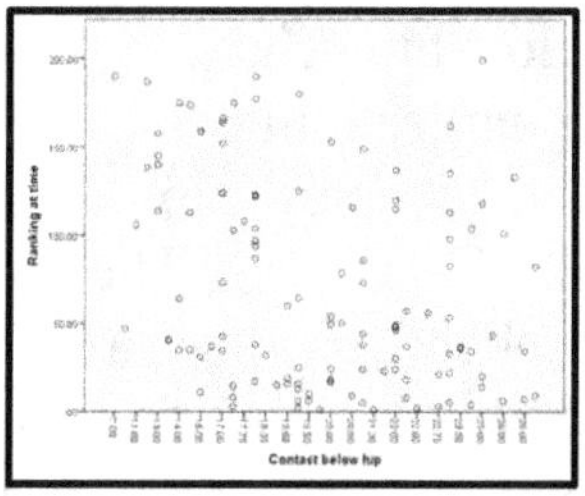

Figure 24 Scatter plot of ranking at time and contact below hip height in the 123 players observed

Height of point of contact below hip height has been found to account for the remainder 19.7 balls out of a possible 30, aligning with the results of Point of Contact. The results show that there is a significant relationship between player ranking and number of height of point of contact below hip height – signalling the more the number of height of point of contact below hip height, the better the ranking.

The relationship between the optimal mixed criteria of deep knee bend and height of point of contact below hip height will now be considered with player rankings.

Ranking and Deep Knee Bend/Contact below Hip Height

Through Mixed Criteria: Knee Bend (deep/slight) and Point of Contact (above/hip/below) it was found that the mixed criteria that consisted of the two most frequent variables, deep knee bend and a height a point of contact below hip height, were the most dominant amongst the world's elite players. The statistics in Ranking have not illustrated a strong correlation between the application of the variables concerned with characteristics of The V by Dr. B ©. It is of interest if this mixed criteria, the last to be considered, identifies a correlation with player rankings. When generating descriptive statistics for player ranking (70.7) and deep knee bend collated with point of contact below hip height, the mean was found to be 17.5 ($SD = 4.5$), aligning with the results of Mixed Criteria: Knee Bend (deep/slight) and Point of Contact (above/hip/below) (see Part 3 for further Statistics). Player rankings were then correlated against deep knee bend collated with point of contact below hip height. Pearson's correlation $r = -0.216$ (one bivariate outlier removed) for deep knee bend collated with point of contact below hip height with $p = 0.008$ statistical significance (2-tailed). The results indicate important scientific implications, and as illustrated in the scatter plot in Figure 25, the contribution found identified that players ranked lower (<50) were reported to have an average frequency between >17 but <22 and players ranked higher (>100) were reported to have a slightly lower average frequency between >15 but <21, when deep knee bend coincides with point of contact below hip height.

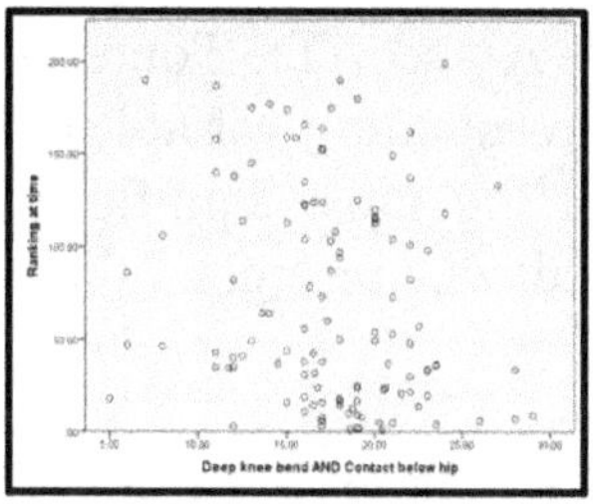

Figure 25 Scatter plot of ranking at time and deep knee bend AND contact below hip in the 123 players observed

It has been identified that not only is this mixed criteria the most dominant, but as illustrated in the scatter plot, players who hold a better ranking (closer to one) are more consistently applying this mixed criteria than lesser ranked players (closer to 200), indicating a significant correlation with player performance. The statistics *unveil* that when deep knee bend is applied with a height of point of contact below hip height, the overall frequency increases in the application of these variables when players are executing their forehand groundstroke. Upon closely considering the scatter plot, it is evident that players holding a ranking closer to one are more frequently and consistently applying these variables.

The overall analysis of player rankings indicated that the average player observed held a ranking of 70.7, and the most frequented variables remained the same as identified in Knee Bend and Point of Contact. When considering the mixed criteria, the results reflected those of Mixed Criteria: Knee Bend (deep/slight) and Point of Contact (above/hip/below).

It was important to identify the average player ranking observed, to ensure a fair distribution of player performances were considered when identifying characteristics of The V by Dr. B © when the playing group (1–200) contacted the forehand groundstroke. It is of interest to more closely inspect players at the top echelon of the game, ranked inside the top 10 on either the WTA or ATP tours. Top 10 Players in the

World on Both the WTA and ATP Tours will now examine the criteria and the corresponding variables in the world's top 10 tennis players, and the most significant mixed criteria, deep knee bend applied when height of point of contact is below hip height.

Top 10 Players in the World on both the WTA and ATP tours

Originally, The V by Dr. B © was *theorised* to be more frequent and consistent in higher ranked players with greater consistency in knee bend and contact at or below hip height. The results in Ranking and the respective correlations were found to neither illustrate the best players in the world, nor if the top ranking bracket (top 10) differed in their application of the characteristics of The V by Dr. B © in contrast to the overall analysis of results.

The intention of Top 10 Players in the World on Both the WTA and ATP Tours is to display the results specific to players ranked inside the top 10 on either the WTA or ATP tours, and to identify if the best players in the world have the same, or a different frequency in contrast to the group as a whole, when applying optimal characteristics of The V by Dr. B ©: deep knee bend and a height of contact below hip height, individually and collaboratively. The analysis consists of 18 participants recorded through the notational analysis process when contacting 30 forehand groundstrokes. The result does not include the top 10 of both the WTA and ATP tours, that is 20 players, due to limitations placed on the observation process (therefore $n = 18$). For a complete data summary on top 10 players in the world and further variables of The V by Dr. B ©, see Part 3 for further Statistics.

Top 10 Players in the World and Deep Knee Bend

The data has revealed that deep knee bend is the most prevalent of the available variables when players use the forehand groundstroke. It has been found that a correlation does exist between ranking and frequency of deep knee bend, but this is only minor. It is of interest to see if the best players in the world are any different from the overall findings thus far.

When generating frequency statistics for the top 10 participants with deep knee bend at point of contact in the forehand groundstroke, the mean was found to be 23.4 (*SD* = 4.0) (see Part 3 for further Statistics). The histogram in Figure 26 illustrates a bell curve depicting that the frequency of deep knee bend in the world's top 10 players is greater than the overall average 23.4 compared with 22.7. In tennis, winning a match at times comes down to merely a handful of points won or loss, and the statistics reflect that the differences between the best players in the world is present, albeit slight.

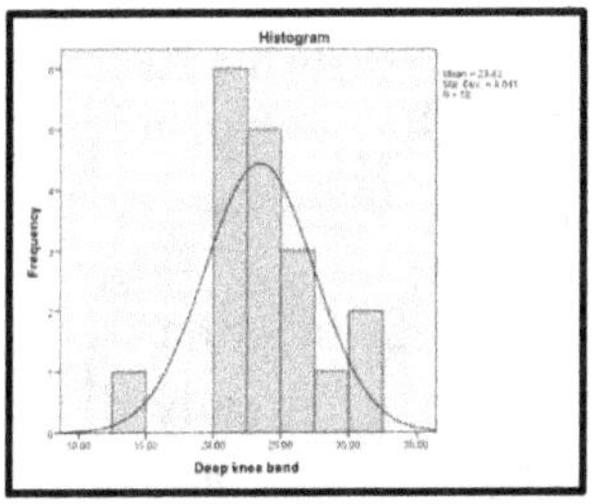

Figure 26 Histogram of deep knee bend in top 10 players

The results are noteworthy, illustrating that there is in fact a difference between the highest ranked players and specifically the elite group of players with rankings inside the top 10 on either the WTA or ATP tours. Players have been found to apply deep knee bend when contacting their forehand groundstroke on average 23.4 times out of a possible 30 balls observed in contrast to the overall mean of 22.7. This result indicates that players who hold a top 10 ranking apply deep knee bend 78% of the time when contacting their forehand groundstroke. The 2% difference between the elite group (top 10) and those ranked outside this ranking range may suggest one reason why the top 10 players are in fact inside this ranking range and are consistently better players than their competitors who hold a lower ranking (outside the top 10).

After all, in tennis, games and/or matches are at times won or lost by only a few points, and this 2% contributes to the differences in the top 10 contrast to the collective player analysis (1 through to 200). The height of point of contact below hip height in the elite group will now be considered.

Top 10 Players in the World and Contact below Hip Height

As indicated in Point of Contact, the variable height of point of contact below hip height has been found to be the most applied height of contact in the forehand groundstroke. As found thus far in Top 10 Players in the World on Both the WTA and ATP Tours, the top 10 players in the world continue to outperform the results of the top 200, suggesting that these differences in performance outcomes may indicate why they are, after all, part of the elite group.

It was of interest to identify if the same frequency or a higher frequency was evident in the already frequented variable. Frequency statistics were generated and the mean was found to be 21.7 ($SD = 3.3$) (see Part 3 for further Statistics). As illustrated in the histogram, the elite group appear to have a higher overall average, whilst some players are found to consistently contact their forehand groundstroke 30 times below hip height out of a possible 30 balls, see Figure 27.

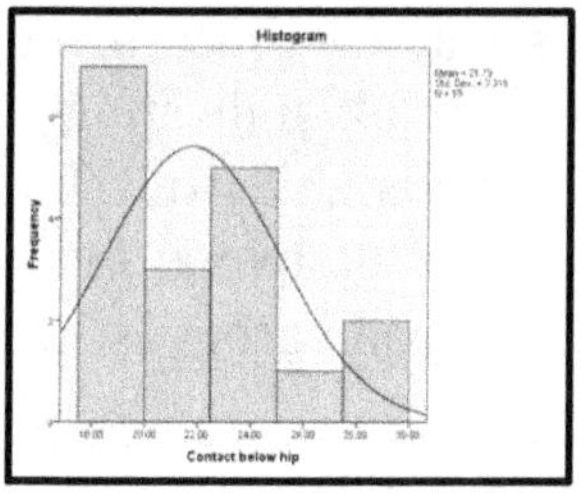

Figure 27 Histogram of contact below hip height in top 10 players

The overall result is important, as it was found that the elite group mean of 21.7 (72%) outweighed the overall average of 19.6 (65%). It has been identified that the 7% difference between the elite playing group and the overall participant group suggests that this is a characteristic used by the elite that may contribute to their ranking status among the best players in the world. The mixed criteria of the two most frequented variables will now be examined, and to identify if the mixed criteria are consistent with the results of Mixed Criteria: Knee Bend (deep/slight) and Point of Contact (above/hip/below).

Top 10 Players in the World and Deep Knee Bend/Contact below Hip Height

When considering the mixed criteria of deep knee bend and height of point of contact below hip height, it was of interest to compare the elite group to the overall mean, found to be 17.6. Given that it was found that the overall participant group applied this combination the most frequently (>58% of the time), it is of interest to identify if this was a defining differential between the elite group and the overall participant group, noting a consistent factor that separates the elite class from the lesser ranked players (outside the top 10).

Frequency statistics were generated and the mean was found to be 20.0 (*SD* = 4.2) (see Part 3 for further Statistics). It was also found, as illustrated in the histogram (Figure 28), that out of the 18 participants, a small portion exceeded the average, with the highest point set at 29 times out of a possible 30 balls. This result is substantial, indicating the elite playing group apply the dominant variables of The V by Dr. B © >66% of the time, exceeding the overall participant group by 8% when contacting the forehand groundstroke.

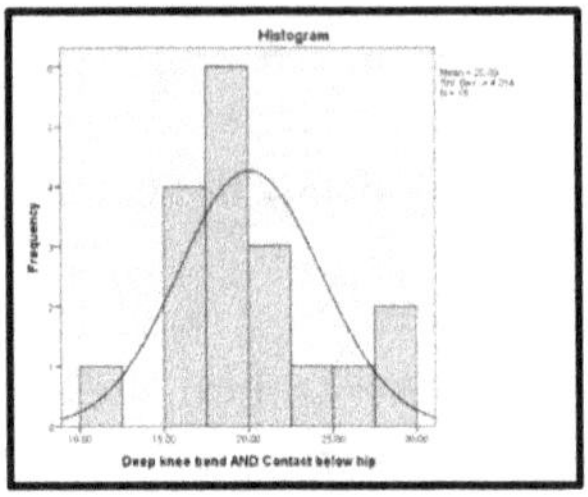

Figure 28 Histogram of deep knee bend and contact below hip height in top 10 players

The findings and resultant *scientific discovery* of the elite playing group, when applying the variables of deep knee bend and height of point of contact below hip height together, not only exhibit The V by Dr. B ©, but outperform the wider participant group.

The results of Top 10 Players in the World on Both the WTA and ATP Tours are substantial enough to reveal the top 10 players in the world consistently apply The V by Dr. B © during the forehand groundstroke, noting that ranking fluctuations between the top 10 occur on a regular basis, and the top 10 analysis offers a truer reflection of player performance.

The results illustrate the performance characteristics of The V by Dr. B ©, as well as those that are not, including the mixed criteria and an overall ranking correlation. In order to corroborate the validity of results and conduct further analysis of the variables representative of The V by Dr. B ©, gender differences need to be considered. Through the observation process, both genders were identified on the WTA (women's) and ATP (men's) tours with the intention of ensuring the results were not one-sided and that an even gender distribution occurred. Given the density of results for the variables of deep knee bend and height of point of contact below hip height, gender differences will now be considered for these variables, singularly and collaboratively, providing a clear overview of the participant group and that not one gender was favoured over the other.

Gender Differences

When The V by Dr. B © was initially *theorised*, it was not gender specific, that is both male and female professional tennis players were hypothesised to apply The V by Dr. B © during the execution of the forehand groundstroke. The intention in Gender Differences is to display the results for the most favoured variables, collaboratively, and to illustrate that irrespective of gender, The V by Dr. B © is applied and present in elite tennis players when executing the forehand groundstroke.

Gender

When observing players throughout the notational analysis, the gender of each player was noted. Throughout the notational analysis there was no way of knowing if an even distribution of male and female tennis players had been notated. A point was made to try and rotate between male and female matches, although the tournament schedule dictated who was able to be observed and when. When looking at the frequency statistics for gender-differences, it was of interest to see if the results thus far were representative of an even spread and that there was not a gender bias. If there was a gender bias, this happened naturally due to the randomly selected matches observed throughout the notational analysis process.

The frequency of male and female participants was 60 males and 63 females, 48.8% and 51.2% respectively (see Part 3 for further Statistics).

The observational process accounted for 60 males and 63 females to make up the 123 players observed and charted against the set criteria. Through the results, although matches were randomly selected, the rotation between male and female matches ensured that the results were representative of both genders and were not biased towards either gender. The results are in fact important, illustrating that the prevalence of the most frequented variables, deep knee bend and height of point of contact below hip height individually and collaboratively, are representative of both genders and do not depict one gender out-performing the other when applying characteristics of The V by Dr. B ©.

All variables of The V by Dr. B © have been considered against gender. It is not the intention of this section to depict each variable and the gender outcome. To see the results of each variable, independent and mixed, the findings have been made available in Part 3. It is, however, the intention to display the findings representative of the most frequent variables, deep knee bend and height of point of contact below hip height collaboratively, and the respective gender frequencies.

Gender and Deep Knee Bend/Contact below Hip Height

Having found through Knee Bend that deep knee bend was the most favoured variable, and through Point of Contact that height of point of contact below hip height is also the most favoured, it was of interest to see if there was a gender difference in The V by Dr. B © and its most optimal variables.

There were 60 male players and 63 female players. The V by Dr. B © is most optimally applied by males and females with an average number of 18.2 (SD = 5.2) and 16.9 (SD = 3.8) respectively. An independent t-test was conducted to see if these two groups were different in The V by Dr. B © and the number of its application. The results (t = 1.6, p = 0.111) suggest that the two groups did not differ statistically (see Part

3 for further Statistics). The scatter plot in Figure 29 illustrates that male tennis players have a consistent range in the application of this position when compared with the female participants.

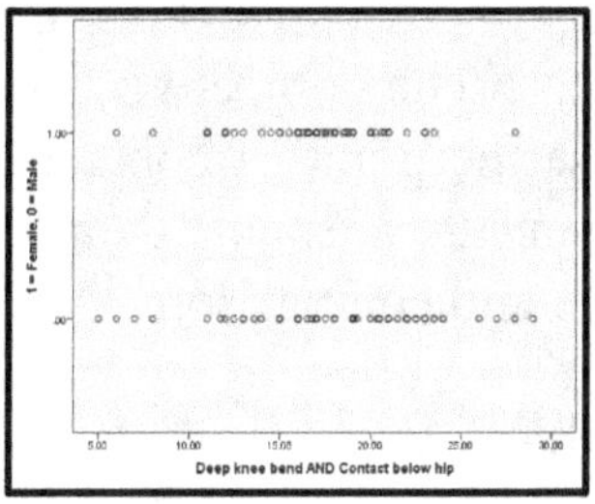

Figure 29 Scatter plot of gender and deep knee bend AND contact below hip height in the 123 players observed

Upon considering gender differences cumulatively, male and female tennis players do not appear to differ in their use of The V by Dr. B ©. Therefore, irrespective of gender, the findings indicate that both women and men apply the characteristics of The V by Dr. B © most optimally, equally, with male tennis players slightly outweighing its application in frequency, yet females identified with a greater range of consistency.

With the cumulative results representative of the participant group, a reasonable ranking distribution (70.7 mean), a near-equal gender share (63 females, 60 males), and a clear share of observational analyses (40.7% each), the overall participant group has been represented in an objective manner and ultimately offers a perspective that indicates elite tennis players and the application of The V by Dr. B ©'s characteristics in the forehand groundstroke.

Conclusion

The notational analysis allowed 123 elite tennis players, ranked between 1 and 200 in the world on either the WTA or ATP tours, to be observed when making contact with their forehand groundstroke. The objective of the notational analysis was to analyse the point of contact, 30 times, in the world's top players, to ensure adequate data could be collected for meaningful analysis and comparison.

The criteria of knee bend was identified to be a dominant factor in elite tennis players at point of contact in the forehand groundstroke. The results found that as a consequence of the variable no knee bend being zero, knee bend was present at all times in the 123 players observed. The frequency of the variable deep knee bend indicated that the lower a player's body position was at point of contact, the more they were representative of the world's best tennis player's forehand groundstroke technique. The criteria of dynamic balance in the forehand groundstroke was found to be a dominant factor at point of contact in the forehand groundstroke, too. This statistic is central, illustrating that the variable balance is a prerequisite for elite players when contacting their forehand groundstroke. The third criteria of The V by Dr. B ©, height of point of contact, found that the traits of The V by Dr. B ©, occurred 90% of the time in players at point of contact in the forehand groundstroke. This suggests that a height of point of contact at/around or below hip height is a common factor among the world's best tennis players at point of contact in the forehand groundstroke.

Player rankings were correlated and a significant but weak correlation was identified between the best players in the world and the use of deep knee bend in the forehand groundstroke. A negative relationship was found between players' ranking and the number of forehand groundstrokes at a height of point of contact below hip height. This suggests that in general the better ranked players hit more forehand groundstrokes below hip height. When considering the mixed criteria The V by Dr. B © and its optimisation was defined, comprising of deep knee bend and a height of point of contact below hip height, a negative relationship was also found with players' ranking. This suggests that in general the better ranked players hit more groundstrokes applying The V by Dr. B ©.

Finally, players who held a top 10 ranking on the WTA or ATP tours were identified to apply the optimal characteristics of The V by Dr. B © more frequently than those outside this ranking range – *forever disclosed.* Irrespective of gender the application of The V by Dr. B © by males and females were not found to be significantly different. Most significantly, the difference in the application of The V by Dr. B © and its optimisation, between the top 200 and the top 10 players, helps explain why only a handful of points can decide the outcome of a match at the elite level which influences player rankings. Further, this finding contributes to understanding the differences in certain players achieving a better player ranking over others, with The V by Dr. B © and its successful *optimal* application reported as a contributing factor to an **elite player status** i.e. a *one-percenter.*

Part 1: Conclusion

The unveiling of The V by Dr. B © is the very *first* of its kind that unravels the answer in tennis – complemented by its transference. Its *three* distinct performance parameters are central to its unfolding as are its respective patterns in application. Significantly, whilst The V by Dr. B © denotes the identified variables, distinctly the criteria differ between the top 10 and players *outside* this niche. Frequency in application is key – the *one-percenters* are denoted as elite *elites* for this very margin that delivers its *scientific* advantage.

The V by Dr. B © is a *collective* optimal performance parameter that holds historical context. Whilst the breadth of scientific literature on sporting techniques in action covers *singular* instances, an equilibrium between action (technique) to reaction (oncoming ball) was void – *yet* to exist. Thus, a compilation of action to reaction and its optimal technical parameters were identified through The V by Dr. B © with its corresponding principles.

It is well known scientifically and at large that every action draws a reaction. To consider an action *without* its correspondence to the reaction removes the autonomy loop – the feedback cycle. Intrinsically, humans are reliant on this correspondence as they cycle through new information on how to react to the next action. Simplistically, it's a decision-making process. By conditioning the autonomy loop on *how* to consistently react has the capacity to cement these optimal performance parameters into the cycle. Autonomy of the cycle delivers the elite player *inside* the top 10 with a more refined feedback cycle then their counterparts – *outside* this range.

The *one-percenters* separate the elite from the elite *elites*. Furthermore, these *one-percenters* have cemented The V by Dr. B © inside their autonomy loop – yet to be integrated into the feedback cycle of the mere elites. However, the feedback cycle isn't without error – the autonomy loop isn't without failure. From time, the correspondence between the action to reaction fails – the player no longer has the capacity to fulfil the reaction to the action. Instances are typically physiological in nature and the player is physically incapable of applying The V by Dr. B © on demand at capacity. A consequence of the failure results in the *one-percenter* losing their ability to perform at this frequency – sliding outside the elite *elites*. The ranking tussle is real and the margins as *The Unveiling* has disclosed, in mere percent's.

Whilst the answer has been unveiled, its application in reality is to come (*Coaching Analyses*) alongside its transference. The technical parameters of The V by Dr. B © are reliant on its respective variables – in unison. Fluidity of movement is key. The V by Dr. B © demands functional movement parameters – the body's natural moveability in action without asserting pressure nor twisting joints that then stretch or increase the applied load. There's an underlying causation for the consistent *one-percenters* – those who are capable of remaining inside this select niche with longevity on their side – the scarcity of injury *afforded* by the functionality of The V by Dr. B ©.

Indeed, there is more than meets the eye. An answer *within* an answer. A historical *first* in its technical compilation that delivers a correspondence between an action to a reaction. An autonomy loop that cements its technical parameters on cue. A scientifically *proven* technical parameter – an *established* optimal performance parameter that adheres to the rule of thumb of human performance – functionality without compromise of performance, a scarcity for inclusion in *coach-education* literature.

Whilst it takes more than The V by Dr. B © to *become* an elite player, it *take*s The V by Dr. B © to *maintain* an elite status and join the *one-percenters*. The embodiment of The V by Dr. B © begins the autonomy loop and further cements the feedback cycle. The uptake of The V by Dr. B © provides a *scientifically established* path to become a *one-percenter* – secondarily, as a direct consequence of its transference aides its *sister-sports* and player success, whilst primarily, tennis players *now* have a scientifically *proven* pathway to attain elite playing success.

PART 2

COACHING ANALYSES

Part 2: Introduction

To identify the prevalence of The V by Dr. B © notational analysis is key. The analytical measures to identify the variables of The V by Dr. B © feed back *into* the confines of notational analysis – indicative collectively in their application at impact. Awareness of the criteria is central to their prevalence (Figure 2). A higher frequency of application *and* transference of the *parameters* of The V by Dr. B © flags a *one-percenter* – herein denoted as an elite *elites*.

The parameters of The V by Dr. B © denote the optimisation of functional movement and its parameters – technical ramifications of the *established scientific pathway* applied extrinsically to the set variables of the criteria, coinciding with *The Optimal Performance Theory* (Berge, 2017). The transference of these technical ramifications is the *real-life* application of The V by Dr. B © in the *real-world* and its implications on player performance. This is disclosed in *Coaching Analyses* for the very *first* time – weighing the current elite calibre players against *not* only The V by Dr. B © but its cemented principle of transference.

As earlier disclosed, gender differences are *null* – with no statistically significant data to indicate *nor* support a gender bias or separation. *Coaching Analyses* follows suit with a more stringent limit – *inside* the top 100 in the world on the respective tours. A consequence of the *reality* of the *Coaching Analyses* is the current active players – at time of publication. *No* player was spared – *all* analyses adhered to the notational analysis confines of *The Unveiling*. *All* analyses were without broadcasting restrictions – publicly accessible. *All* analyses were unknown and randomised to prevent bias. *All* analyses stopped once the total player-share aligned with *The Unveiling* – 51% female, 49% male. The compiled data is representative of accessible players *scientifically* analysed for the very *first* time against the principles of The V by Dr. B © falling inside the controlled constraints. Noteworthy, *three* male players have been included despite their ranking outside the top 100 – as *former* one-percenters. Recall, a consequence of failed results is the *one-percenter* losing their ability to perform at the frequency upheld by The V by Dr. B © – sliding outside the elite *elites*.

Scientifically, this is a *first* to be disclosed. Pragmatically, it's The V by Dr. B © in *reality* – applying transference. Simplistically, a four-way division was primarily used to segment a player's game prior to the scientific analysis – forehand *and* backhand groundstrokes, service motion, return of serve, and (angular) movement. *All* other aspects of the game derive from these divisions i.e. the overhead (serve), the approach (forehand), and the volley (*fore*hand, serve and dependency on movement) negated as a result.

> Recall, **The Rule of Transference**: *a functional application of one movement and its technical parameters are inevitably transferred as a consequence of conditioning, whereby a player is conditioned to perform a set task accomplished through the sound execution of the conditioned technique and its set discrete skills. Thereby, these skills are inevitably tied to the players rudimental skills of technical performance parameters and their body's functionality to perform them. Thus, The Rule of Transference is deeply rooted in the players conditioning of a skill and inevitably its technique. With likened properties, this skill is thus transferred in its application and through its very conditioning, aiding a player's ability to perform a set or given technique against its predecessor – contributing in large to the output of optimal performance parameters.*

Numerically ordered as per player ranking *pre*-analysis, both male *and* female players are listed with disregard for gender as per its null statistical inference. The *Coaching Analyses* discloses the *scientific* implications of a player's technical parameters – systematically. Failures of The V by Dr. B © and its transference are denoted by a player's prospects as a *one-percenter* in action, whilst amplifying the prevalence of the respective variables, dismantled, *before* identifying their correlation to player ranking success. The *Coaching Analyses* disclosed is the emergence of The V by Dr. B © coach-ationally *and* scientifically to further unveil **What is Your Game Missing?** and its applicability across the playing sphere, beginning with the World Number 1's at time of analysis.

(Ranking, Pre: 1) Simona Halep

The current leader of the women's tour, this player has consistency of performance on her side with an increased frequency of optimal performance parameters. When serving the player begins by activating her lower body which is coupled with the players upper body rotation. Whilst the player has an optimal reach as she drives up to her ball toss, she is slightly held back by the push in her lower body – greater flexion would result in the player with a more upwards lift and resultant momentum post impact. Yet, the player has consistency on her side as the arch in her back is offset by her lower body – both slight, working with the player, however there is room for an increased range of movement. Further, the player exhibits flexion at her hips pre and post impact – an optimal performance indicator, that helps her compensate for her limited momentum at times through the ball. At the top of the elite *elites*, consistency of play is the best winner in the players performance.

Upon return the player has slight lower body activation at times which halts her ability to extend through her returns. The player does however lean her body, upper and lower, into the return, that sees her effective use of kinetic energy to counter the pace of the serve. With a controlled proximity to the ball, more often out in front, at times the player gets caught late when the pace of serve speeds up. Primarily, the players controlled height of point of contact optimises her return. Whilst the depth in her range of motion could further optimise her return, the players current consistency in her activated lower body, state of balance and height of impact set her aside from the significant majority.

The player flexes her lower body through her groundstrokes as she exhibits optimal performance parameters off both sides. Through impact the player manages to release torque through her hips which is furthered by her use of angular positioning. With an open stance off both sides, the players upper body rotation stays in sync with her lower body flexion to not only release her body as a whole behind the ball, but equally the generated kinetic and elastic energy. The players release through impact is compounded through her use of torque as she transitions through, whilst further use of her extension through and into the court would see the elite *elites* further capitalise on her groundstrokes.

A mover that exhibits optimal performance parameters throughout her game, the player integrates vertical, angular and horizontal movement throughout. With the consistent presence of dynamic balance the players level of control is optimised. Although slight, the player tends to slow when her momentum through the ball is halted, whilst similarly a result of her extension through the ball – at times steady when she isn't the one dictating play. Ultimately, the player is an elite *elites* with flaws present when her consistency wavers, which is limited. A key optimal performance parameter is her application throughout of her lower body and controlled height of point of contact in sync – this consistent application alone is rare and a performance parameter that sets her apart from the majority of fellow tour players.

(Ranking, Pre: 1) Novak Djokovic

At the current echelon of the game, this player not only frequently exhibits optimal performance parameters throughout his game, but his consistent and near unwavering level of dynamic balance allows him to excel almost regardless of his court position. Through his serve the player generates ample lower body flexion that he uses to compound the arch in his back, using his coil-like technique to generate torque to release through as he extends through his ball. The player achieves optimal momentum into the court complemented by his flexion at the hips that optimises his transfer of elastic and kinetic energy through his serve. The players serve is further optimised with the location of his ball toss, working with his body – a clear optimal performance parameter as a whole delivered by the elite of the elites.

As the player returns serve, he keeps a deep level of flexion in his lower body as he impacts the ball out in front and at the optimal height of point of contact as he often extends through his return. The players return is further complemented by the flexion present at his hips, more often than not, that allows the player to transition into and through the ball. Whilst there are times the player is placed under pressure and his optimal performance parameters are compromised as a result of his opponent, most frequently the player dictates play courtesy of his key performance indicators that he consistently executes through his return.

The players forehand and backhand work in sync near seamlessly with his body as he achieves a constant lowered centre of gravity with deep flexion through his knees and through impact, coupled with his optimised height of point of contact as he cleanly extends through the ball – limited to the elite *elites* and showcased by the current world's leading male player. Further, not only does the player consistently apply the optimal performance parameters through his groundstrokes, he transitions through his shots through to extension, before transitioning to the next ball – an elite *elites* performance indicator as a result of his speed of movement and shot anticipation, limited to the elite niche.

Court movement for this player is at the pinnacle of the game. Not only is angular movement integrated firmly into his game, but his anticipation of the ball and his setting prior to contact ensures he's able to achieve ample torque through his midsection as he positions his hips angularly in position to the court, allowing him to near-flow from point of contact, to the next point of contact. Courtesy of the players consistent presence of dynamic balance, and his activated lower body with a deep level of flexion, this recipe is seldom found in the elite, and near exclusive to the elite *elites*.

Overall, there are a limited number of players on both tours who are capable of applying a seamless integration of optimal performance parameters, as with extension through impact and the use of angular positioning. As the world's current leader on the men's tour, it is no surprise that this player ticks all the boxes. An area for further development for this player would be tied to his height of point of contact on his backhand – at times the player is caught standing taller than most optimal coupled with a higher point of contact that often either forces the error from the player, or is unforced. Keeping a consistent lowered state, as with the ability to more readily move up to these balls as to control its height of contact, would be key to conquer this slight weakness.

(Ranking, Pre: 2) Angelique Kerber

This is an elite all round player with minor flaws to address. The overwhelmingly exceptional use of her lower body in her groundstrokes allows her to compete against the best in the world – and deliver. The lower body mechanics are in sync whereas flaws creep through when her lower body isn't switched on – rarely the case. However, her rotation is opened up through the use of her lower body – transitioned through her shots. This combination is often overlooked throughout the ranks and typically key of players inside the top 10 and more frequented by players who reach the top on either tour.

Return of serve is a weakness when refusing to move to the ball, or position herself slightly backwards. By standing her ground the player is unable to control her height of contact placing increased pressure, and reliance on her upper body. To the contrary, by controlling her height the already in sync mechanics of her lower body would enable her return to be near unstoppable – a dangerous shot to add to any players arsenal. At this level of play it is the very minute changes, as this player exemplifies, placing a heavy load of reliance on her vertical movement – the powerful role of momentum.

When addressing the serve the player has a solid technique – upwards lift is present as too is momentum. Yet, the depth of the lower body has range to increase that would allow the player a slight increase in her range for propulsion and force applied at impact. Second to this is her upper body rotation – there is room to increase the degree of separation to "wind up" and to release into her point of contact.

The players groundstrokes are solid on both sides. Her horizontal movement is a strength, meanwhile although she doesn't consistently cut off the angle of oncoming balls, she can do so when her movement has found its rhythm. The players quickness to react vertically through rallies isn't as strong as her horizontal gains – this is why it's more evident to highlight, although merely slight. By working towards a deeper integration of angular movement into her game through further development of speed of reaction would see this player close to unbeatable.

(Ranking, Pre: 2) Rafael Nadal

When serving the player achieves optimal lower body activation in line with his upper body rotation that not only achieves optimal torque, but also drops his elbow to reach towards the ball. A result of the players optimal lower body flexion sees his ease in reach towards the ball compounded by his ball toss located inside the court to help the player transfer not only his kinetic energy, but to achieve momentum inside the court. As an elite *elites*, the player exhibits optimal serving parameters that not only maintain the use of dynamic balance throughout, but effectively relishes in his lower and upper body in sync throughout his service motion.

Upon return the player maintains a lowered centre of gravity with an optimal forwards lean as he flexes at his hips. Further, a result of the player already positioned to move with the ball, sees the player extend through his return accompanied by his use of momentum. The player uses his technical parameters to his advantage as his body moves with the ball, whilst further, as a result of achieving an optimal height of point of contact, the body's mechanics are at ease as the player merely transitions through his shots. As a elite *elites*, the players ability to return is second to none as with his use or angular positioning to gain further advantage.

The players consistent presence of dynamic balance throughout his groundstrokes not only allows him an increased level of control through his shots, but further complements his use of angular positioning. As the player achieves a lowered centre of gravity throughout – deeper than the majority, a key parameter of the elite *elites* at point of contact, tied to his lowered height of point of contact, this player exemplifies the elite niche. Interestingly, the player wavers in his performance as his proximity to the ball, as with his height of point of contact falls out of sync. Although this is scarce, the players positioning is compromised when confronted with higher balls or balls caught slightly closer than most optimal – unable to extend through his shot.

With the players court movement of the elite *elites* – horizontal, vertical and angular movements are applied with ease and near seamless. Whilst the player has a preference for baseline play, he readily moves vertically as he adjusts his proximity to, or away from, the oncoming ball. Most significantly, the players ease and use of his angular positioning is perhaps his greatest strength as he is able to rotate through his shots, placing him in a position to extend through his groundstrokes that in turn primarily puts the player in an offensive position. Yet, perhaps the players greatest weakness is at times his reliance on his upper body – used to take the load when the player catches the ball slightly close to his body, or high, that then inhibits his lower body from taking the load. However, this comes with its own catch – allowing the player to get out of 'sticky' situations, yet his body is placed under a higher load than most optimal, placing increased pressure on his joints. This is why performing with optimal performance parameters consistently is a must at the elite level as it disperses the load and places it where *it is* technically and functionally most able – something this player has more often than not been able to accomplish throughout his career.

(Ranking, Pre: 3) Roger Federer

The functional applications of this players game is why longevity has been in his favour on the court at the highest of levels – consistently. The functionality in itself has been responsible for his smooth technical applications that a generation has near dreamed of replicating, yet fallen short. This player is the *echelon* of the game, with angular movement firmly cemented inside his game, while as the years have passed, slight creases find their way into his near-flawless game. A primary one-percenter key remains, coinciding with balance, the players integration of angular play embedded in *all* areas of his game – scarcely found elsewhere, denoted as the newest optimal performance parameter to enter play.

A service motion of lower body activation, hip rotation and torque, places heightened control on the upper body opposed to limitations – the functionality is how the body was and is *designed* to be. Upper body power is a given, yet is not of reliance due to the effectiveness of their lower body. The same applies for momentum – it is easily achieved through the control of his centre of gravity at all times. This all-round control permits the player to fall into the court, recover, or even at times head to the net with astounding speed – courtesy of his propulsion into the court post service impact.

Groundstrokes that make a scientist's heart sing. His lower body remains in constant activation – a rarity often only found in a mere handful of the elite. Although comfort for some, this isn't without waiver with at times this sliding higher and placing him at greater jeopardy with his opponents – the ever so slight gap here is often where errors fair. This ever so slight area is well hidden, mostly, through his state of balance – a constant courtesy of his centre of gravity constantly under control. This is further complemented by his upper body rotation on both sides, utilising his stored elastic energy with each and every shot. The location of contact is key – out in front, just within reach, permitting the smoothest of transitions for the racket strings to glide across the back of the ball.

Offensive or defensive, the players balance is undisputed with a fall-back effort supported by the relocation of his centre tee of gravity – pushed forward upon contact, disallowing his opponent to force him entirely on the defensive. Recovery here is key and more readily achieved through his constant state of dynamic balance.

The return differs ever so slightly with his at times sliced "block" return. An element not too many can get away with, this player manages with controlled finesse to position his return accordingly. Commonly a technique of this type used on the return places the receiver on the back foot – the angle of his racket face against the back of the ball when blocking, or bunting, is rather an uncomfortable shot to be dealt as a server. Lower body mechanics are key – activation is identifiable as with his groundstrokes, as is the location of his return – controlling his height of contact, and location, with minor placement positioning. Forwards momentum into the return isn't as consistent as his groundstrokes, the slightest of areas for further consideration.

(Ranking, Pre: 3) Caroline Wozniaki

Through her serve the player generates optimal extension as a result of her inside court ball toss. By dropping her lower body in sync with her elbow drop, she extends through her lower body in alignment with her upwards drive to optimise the release of both her kinetic and elastic energy. Exhibiting optimal performance parameters through her serve, the player showcases why she's inside the elite *elites*. At times the player's elbow drop is not as pronounced, limiting the player's ability to generate optimal elastic energy. By dropping her elbow slightly more, as with leaning her inside hip towards the court, this slight advancement will contribute to further optimising the player's serve.

As the player returns she exhibits an activated lower body tied to an optimal height of point of contact as she flexes at her hips – leaning into the return. Not only does the player exhibit optimal performance parameters through her return, as an elite *elites* this is no surprise. Yet, from time the player does catch the ball higher than optimal which limits her transition of kinetic energy through the ball as she lifts through impact. Although slight, this inconsistency often places the player on the defensive in need of scrambling to regather lost ground. The player uses an abbreviated swing at times through her return – an optimal performance indicator, whilst often extending through impact. With slight limitations in the players consistency, this is key to solidify her position in the ranks, primarily consistency in height and depth respectively through her return.

The player achieves optimal performance parameters consistently through both her backhand and forehand as she maintains a deep level of flexion through impact, whilst extending through the ball, complemented by her consistent optimal height of point of contact. Not only does the player extend through impact, she often uses her momentum behind the ball, finishing forward, or releasing through her hips to finish facing the direction of her ball. With good rotation through her upper body, the player often transitions through her shot. When on the defensive, the player has the tendency to catch the ball slightly close to her body – as a consequence she's reliant on her upper body taking the load of the shot in contrast to her lower body. Further development on her proximity to the ball, and lower body depth when on the defensive and her consistency of application, are key for the player to keep her hold inside the elite *elites*.

With movement as the stronghold of her game, the player uses her agility to her advantage consistently through play. With the use of horizontal, vertical and angular movement, the player exhibits optimal performance parameters as a direct result. Further, the player adjusts prior to contact to ensure her height of point of contact is most optimal, although when on the defensive the player does remain at times higher than most optimal with the tendency to overstep her location to the ball. However, the players angular positioning is a decisive factor of the elite *elites* that enables her to take time away from her opponents whilst placing herself in a position to either dictate play, or continuously place her in a position to run down her opponents' ball as a result of angular play integrated into her game.

(Ranking, Pre: 4) Alexander Zverev

Lower body mechanics of the player throughout his groundstrokes are in sync and his consistent knee bend is a performance parameter that sees this player as part of the elite *elites*. Most considerably, the players lowered base permits a heightened state of balance that then allows the player to smoothly reach and extend through his groundstrokes opposed to being pushed around – further compounded by his use of angular momentum. This state of play is limited to the elite *elites* for good reason – a serial skill comprising of multiple discrete skills, complicated in its execution when angular play is integrated and the players ability to then 'smooth through impact'.

The players service motion has a heightened reach and sound rotation, yet slightly lacking in the upwards lift. Although the lower body is activated throughout the serve, there is depth to be gained to further propel the player into the court. Whilst momentum is present, a feat in itself with his ability to recover back to the baseline, a sturdier base is key as with an increased range of movement. Height of contact coupled with the location of the ball at contact are key performance parameters. This player has captured this key and likewise applies by ensuring he's "just" within reach of the ball upon contact to ensure a smooth transition through the ball. This is often overlooked as poor positioning whilst rather it is a key elite performance indicator by way of applying balance and lower body positioning to have the greatest impact upon contact – added control of the ball.

Return of serve has room for further development. Although the players groundstrokes notably use the lower body's mechanics, upon return he sits taller than most optimal, causing his height of contact to be greater than desired. This is a notable weakness exhibited by the player, however likely only advantageous to the elite *elites* due to his already smooth transition to baseline play.

(Ranking, Pre: 4) Elina Svitolina

When serving the player slightly activates her lower body prior to extending upwards and through the ball at impact, with a slight arch in her back that allows the player to generate torque prior to releasing through impact. Whilst at times the player's ball toss does land inside the court, this varies in consistency. However, the player is limited in the depth of her lower body that contributes to her at times restriction in her service speed – easily addressed with greater depth and ultimately an increase in the players stored kinetic energy. A technically clean serve, the players primary hold back is her lower body that otherwise keeps her inside the elite *elites*.

As the player returns she flexes slightly at her hips as with a slightly lowered centre of gravity that begins to drop at and through impact to exhibit deep lower body flexion more often than not. The player frequently extends through impact as she also transitions through the return in contrast to pulling off the ball. Courtesy of the players optimal performance parameters, her noted weakness is in her consistency of depth at times in her lower body that in turn affects her height of point of contact. However, most frequently the player exhibits a technically smooth and controlled return that identifies her as an elite *elites*.

The players groundstrokes not only consistently exhibit the presence of dynamic balance, but most significantly the integration of angular movement. Not only does the player keep a lowered centre of gravity with often deep lower body flexion through impact, likewise her height of point of contact is primarily consistent throughout, exhibiting optimal performance parameters through both her forehand and backhand. Further, the player manages to transition through her shots at impact and extend through the ball, whilst also often allowing her momentum to take her through impact in contrast to holding back – courtesy of her angular positioning through play. Showcasing groundstrokes of the elite *elites*, the player's current weakness is merely in her consistency of impact at times that places her in a defensive position. To strengthen her game, consistency of height of impact is key for this player to build an even stronger offensive game.

With movement a strength for the player, this is compounded by her use of angular movement that brings the players game to the next level. Whilst horizontal movement is second nature, as with her vertical movement as she adjusts to the height of the ball, her ability to position her hips inside the court, cutting off the angle of the ball whilst taking time away from her opponents is a leading performance indicator that maintains her game as a elite *elites*.

(Ranking, Pre: 5) Naomi Osaka

The players speed of rotation through contact of her serve is a clear optimal performance indicator. Whilst her lower body flexion is limited, as with her torque, the location of her ball toss inside the court is enough to generate upwards lift to the ball. Limited in her propulsion and upwards drive, momentum is key as with her simplified rotation in the effectiveness of her serve against her opponents.

Upon return of serve the player remains inside the baseline accompanied by an activated lower body through impact. At times the player moves with the ball at contact, exhibiting momentum further inside the court, taking advantage of her stored energy. Whilst the lower body is activated, most notable is the player's ability to extend through her returns more often than not – an optimal performance indicator.

Not afraid to step into the court when contacting her groundstrokes, this is perhaps the player's greatest strength paired with her solid lower base of support and use of dynamic balance. The player contacts both her forehand and backhand at the optimal height of impact most frequently, whilst her upper body rotation is key in her shot outcome allowing the player to hit all the way through the ball. At times the player's extension is limited, although slight, which is tied to her proximity to the ball, which would allow her even greater control and momentum integrated into her groundstrokes.

The player slightly catches her backhand at times outside the optimal location of impact, whilst her forehand falls short of finishing her extension through the ball. However, the players all-court movement, horizontal and vertical, whilst not being shy to move into the court, allows her momentum to be a key elite performance parameter. Likewise, her base of support paired with her optimal height of contact more often than not, can progress with a heightened integration of angular movement to allow the player to further take time away from her opponents and maintain a consistent elite *elites* performance.

(Ranking, Pre: 6) Sloanne Stephens

When serving the player activates her lower body which is then offset by an arch in her back before the player reaches upwards to begin her lift to the ball. At times the player exhibits an inconsistent ball toss, slightly closer to her body, that restricts her forwards momentum. As an elite *elites*, this window of inconsistency is an opportunity of few that opponents readily take advantage. However, when the player positions her ball slightly inside the court her technique begins to exemplify her optimal performance parameters, coupled with deeper flexion in her legs and a more inside court positioning post impact. From the players overall technique, optimal performance indicators are present, yet a result of the player's mostly consistent and optimal service motion, when her ball toss and legs are slightly out of sync, the player's outcome is ultimately affected.

Upon return the player is slightly flexed at her knees before lowering her centre of gravity more at contact as she drives through the ball. The player rotates her upper body smoothly through return as she leans into the ball, whilst when anticipating the return most optimally, the player will use her angular positioning to cut off the balls angle – an optimal performance indicator. The player uses an abbreviated swing on return and more often tends to limit her extension through the ball which is paired with the restricted use of momentum. Yet, this is conditional to the players objective, as when the player chooses she does in fact step up and extend through, taking a more offensive stance – an optimal performance indicator. Whilst inconsistencies are present on the players return, this equally affords a player at this level to keep her opponents on their toes. Interestingly, contributing to the player's controlled height of point of contact, dynamic balance and activated lower body, rather than being limited in her restriction of movement through the ball, she uses this time to prepare for the next ball which more often than not is set with a deeper range of motion and use of angular positioning.

The player keeps a low centre of gravity with a deep range of motion through her groundstrokes – an optimal performance indicator. Likewise, when contacting her backhand the player rotates through her upper body with a height of point of contact most optimal. Similarly, off the players forehand she exhibits a height of point of contact of the elite *elites* paired with her activated lower body and consistent presence of dynamic balance – an optimal performance parameter. The player tends to utilise her the torque generated in her body when contacting the ball opposed to applying momentum – although slightly limiting, the players consistent presence of dynamic balance, paired with her upper and lower body in sync through her point of contact, affords the player to keep her positioning that is often limiting when this level of ease at impact and dynamic balance is scarcely achieved.

With a preference to maintain her baseline positioning, the player is comfortable horizontally more often than not. Although the player tends to stick to the back of the court, she rarely moves further back, keeping her ground. Yet, the player does move vertically to ensure she meets the ball at the optimal height of point of contact. Similarly, the player utilises her angular movement to her advantage – an optimal performance indicator. By opening her hips to the court whilst maintaining an angular stance, the player exhibits qualities of the elite *elites* as she continues to use the fluidity of her technique to extend through the ball. Although inconsistencies remain, the player is most dominant when she is consistent that it is dependent on her opponents pressing her inconsistent patterns to take advantage of a lull in performance often countered by the player as a result of her presence of optimal performance parameters.

(Ranking, Pre: 6) Kevin Anderson

When returning the player activates his lower body as his upper body leans slightly into the court. The player rotates his upper body as he uses an abbreviates swing – an optimal performance indicator, with slight momentum behind the ball. Further, the player frequently extends through his return, yet from time does get caught too late – a result most often from his proximity to the ball. To increase his return effectiveness, angular positioning as with the players anticipation of the ball, will allow the player to readjust his positioning ever so slightly to optimise his already applied optimal parameters of his lowered centre of gravity and optimal height of point of contact achieved through his returns.

Through his serve the player activates his lower with an optimal extension upwards and through the ball. Not only does the player manage to keep his lower body in sync with his upper body, the player's ability to extend and drive upwards in unison, further compounds the optimal location of the player's ball toss. Near optimal in his service application, the player's slight weakness comes from times as he either hits down, too early, or through – catching the ball slightly late; therefore, minor timing of his upper and lower body in sync, is slightly off. All in all, this player is an elite *elites* with a serve that is an optimal performance parameter.

The player keeps a lowered centre of gravity through his groundstrokes as with the consistent presence of dynamic balance – an optimal performance indicator. Through the players forehand and backhand he does exhibit flexion at the hips as he leans into the court, yet his momentum behind the ball at times is often scarce. As a elite *elites*, the player's frequent extension through the ball is key, as with his use of angular positioning – although there is room for a higher level of consistency. The player keeps an optimal height of point of contact through his groundstrokes most often – a key aspect of his performance. As the player manages to maintain his flexion in his lower body and control his height of point of contact, these are primary facets of the elite *elites* and optimal performance parameters for this niche.

As the player positions himself throughout play, he predominantly exhibits horizontal movement through play, whilst he does integrate angular movement too – an optimal performance indicator. The player isn't afraid to move vertically into the court, yet opts to stay back most often, where his biggest winner is in his ability to use his angular positioning whilst keeping a lowered centre of gravity and momentum behind his shots more often than not. To progress his movement the player's anticipation of the ball, as with his agility to then set and reduce the opponents' angle, would be key factors to take his movement further.

(Ranking, Pre: 7) Petra Kvitova

When serving the player flexes her lower body in sync with the rotation of her upper body before driving upwards and rotating out through the ball. Exhibiting optimal performance indicators of the elite *elites*, the player not only achieves momentum behind her serve, but is completed by a ball toss inside the court. Although indicative of the elite *elites*, the player is limited slightly of her projection behind the ball that results in a held back follow-through at times. Whilst merely slight, equally, a more pronounced presence of flexion in the players lower body would see her release of kinetic energy increase to support a more upwards, and outwards, release behind the ball.

Upon return the player manages to maintain a lowered centre of gravity which contributes to her balanced positioning. Flexing at the hips results in the player's upper body leaning into the return, whilst her height of point of contact is predominantly consistent with the player not shy of moving towards, or away from the ball to adjust the height – an optimal performance indicator of the elite *elites*. The player rotates her upper body through her returns, however abbreviates her take back – a more direct line of her racket opposed to her circular motion used on her groundstrokes. This behaviour of the player and her timeliness to react to the return is an optimal performance indicator – adjusting her game to ensure she maintains a lowered centre of gravity coupled with a controlled contact height inside the optimal range of impact. With momentum transferred behind the ball, this player is a elite *elites*, whilst to further progress, a more consistent use of her angular movement through her returns would see her dictate play from the get go.

The player maintains a lowered centre of gravity as with the consistent presence of dynamic balance through her groundstrokes. With an activated lower body pre and post impact as with her upper body rotation, the player continues to exhibit optimal performance parameters. The player manages to control her height of point of contact courtesy of her proximity to the ball and consistently activated lower body. Although there are times when the player is pushed back, she still manages to maintain her positioning courtesy of her ability to slightly move inwards, backwards, or outwards to adapt to the oncoming ball whilst keeping an activated lower body. At times the player holds back on her extension through her backhand whilst making contact with the ball with slight lower body activation opposed to the more pronounced depth of the elite *elites* – a slight increase in consistency here will move this player inside this select niche.

With movement on her side, the player applies horizontal, vertical and most significantly angular movement to her advantage through play. A result of the players use of angular positioning allows the player to set her proximity to the ball prior to impact, whilst increasing her ability to rotate through her hips at impact, opposed to a more horizontal position that limits the use of this technical parameter. Further, the player more often than not uses momentum through her shots, however this is an area for the player to further progress – at times holding back on her forwards transition post impact which limits the player from maintaining her offensive position.

(Ranking, Pre: 7) Marin Cilic

A pronounced level of lower body flexion is apparent through the players serve that allows him to use his momentum behind the ball. As the player lowers himself to the ground he optimises his longer levers as he extends upwards towards the ball. Courtesy of the players arch that generates torque through his midsection, as with his inside hip sway into the court, the player releases through impact to optimise his transfer of energy. The player is limited at times of his use of momentum as he can be caught landing slightly in front of the baseline post-impact – holding back on his open release through the ball. A more optimal release through impact will look to advance the player's serve even further.

As the player returns he keep a deep level of lower body flexion – an optimal performance parameter and indicative of the elite *elites* as he achieves an optimal height of point of contact as with flexion at his hips whilst he extends through impact. As a elite *elites*, it's of no surprise the player's technical parameters are in sync through play, as with the consistent presence of dynamic balance. The player's ability to not only keep his centre of gravity low, but to also extend through whilst keeping his upper body in sync, are further complemented by the players use of angular positioning through his return – an optimal performance indicator. If the player is able to lower himself slightly further through impact and extend on his momentum, he will solidify his return game inside the elite niche.

The player exhibits optimal performance parameters through his groundstrokes – an identifiable characteristic of the elite *elites*. Not only does he use momentum to his advantage, he optimises his release of elastic energy as he primarily extends through impact off both sides. Whilst in motion, the player equally is able to maintain the presence of dynamic balance as he transitions through both his backhand and forehand as he extends through – an optimal performance indicator. With the use of angular positioning, the player optimises his groundstrokes even further – privy often to the elite niche. Yet, the player does from time tend to be positioned further behind the baseline than ideal, placing him in a defensive position. To further optimise his game, stepping up to the baseline more often, whilst advancing his anticipation to the ball, will see the player climb to newer heights inside the elite niche.

Courtesy of the players angular movement he is readily able to move around the court as he anticipates the oncoming ball. Whilst horizontal movement is common with his baseline position, he often moves vertically as he closes off the point. However, the player also moves vertically as he backtracks behind the baseline at times to control his height of point of contact – an optimal performance indicator. Yet, the player gives himself the advantage when he uses his angular positioning to further release through impact and extend. As a elite *elites* the player moves smoothly and most often, with fluidity around the court. An area to further progress is the players proximity to the ball as he applies his angular movement – more readily stepping up to cut the angle of play off, as with at times a slightly lowered centre of gravity.

(Ranking, Pre: 8) Karolina Pliskova

Although a player that has had tremendous success with her serve over the years, her lower body drive leaves a lot of room for progression with limited depth achieved during her service motion. This is further compounded by the torque restriction that is relayed to the upper body and its slight rotation. The player does offset these limitations with her smooth motion and rather optimal ball toss location – setting her up for an easier flow into the court which benefits from her stored elastic energy. Meanwhile, the player does have solid control over her centre of gravity which allows her service motion to prosper despite its limitations.

Upon service return the player activates her lower body, however rises before contact – neglecting to use her stored energy against the oncoming ball. The height of contact is then also compromised, yet the player is in tune with her vertical movement and isn't afraid to move into the court post contact – keeping momentum on her side. This is an optimal return action which would be compounded with a more active lower body pre-impact.

The players groundstrokes are effective to say the least, however most significantly is the height and location of her point of contact. Although she opts not to optimally activate her lower body, her centre of gravity remains controlled resulting in a constant state of dynamic balance. Interestingly, the player achieves good upper body rotation through her groundstrokes, on both sides, with the gap present in the use of her lower body. It becomes clear this reduced use of her lower body is tied to her vertical movement during play – being more comfortable waiting for the ball to come to her, opposed to moving up to the ball and better controlling the height of contact. A direct consequence of this is the player missing out on the benefits of her lower body. To the contrary, this player is an elite of the elite and these limitations are merely slight. Further, at times of need the player does activate her lower body as desired which leaves the question open to consistency of play and timing as key performance indicators to reference.

A result of the players longer levers than most players is her ability to cover more ground in a shorter time frame. The players speed is complemented by her optimal reaction time, utilising her natural strengths to her advantage. These strengths tied in with her controlled centre of gravity keeps her inside the elite *elites* – addressing her lower body use in serve, return and groundstroke consistency would see this climb even higher.

(Ranking, Pre: 8) Dominic Thiem

When serving the player generates deep lower body flexion that allows him to drive up towards his ball. Similarly, the player keeps his centre of gravity in alignment with his core as he arches his back to level his knee flexion before releasing his torque upwards through his extension as he release through his hips – an optimal performance parameter. Although the player's lower and upper body are in sync as he showcases a clean technique, his ball toss hinders his momentum forwards into the court – a direct consequence of his reduced service speed. With the players effective mechanics, a slight shift in his ball toss as with his extension more outwards rather than brushing the back of the ball, will help him further optimise his serve.

The player achieves an activated lower body through his return whilst he often keeps an optimal height of point of contact. Whilst the player exhibits flexion at the hips through his return, he also manages to extend – an optimal performance indicator. As the player transitions through his return, he rotates through his upper body with often an abbreviated return. Not shy of stepping through his return to use momentum or his technical dynamics to his advantage, the player's weakness is when he lifts through impact, often a consequence of the player not achieving a low enough centre of gravity prior to impact.

Through his groundstrokes the player keeps an activated lower body as with his consistent presence of dynamic balance. Off his forehand the player maintains an optimal height of point of contact as he often manages to transition through impact, however is caught from time limited in his momentum behind the ball. Similarly, the player's one handed backhand relies on his lowered height of point of contact, demanding flexion in the lower body. Yet, the player does have a tendency to exhibit slight flexion through impact in contrast to lowering himself further as the elite *elites*. When playing from the front, the player does manage to maintain optimal performance parameters for his groundstrokes – with consistency of application noted as his weakness. Whilst the player sways between his resultant technical application of the elites and with his higher centre of gravity, a more consistently pronounced level of flexion in his legs more dominantly will allow the player to better optimise his extension through impact and his open release through the ball. Primarily, it starts with a wider base of support and once achieved consistently, will see the player take advantage of his elite technical assets.

Primarily the player maintains a baseline presence with the tendency to stick behind or on top of the baseline, moving vertically forward when the shorter ball approaches. With horizontal movement coming with ease for the player, this is further optimised through his clear and decisive integration of angular movement. Not only does the player manage to integrate angular positioning too, his ability to generate torque and unwind through impact whilst directionally changing the flight of the ball is not only an elite performance parameter, but a decisive performance indicator that allows him to maintain his elite niche when his level of consistency wavers. To look at solidifying his position inside the elites, the player is in need of greater consistency through play, keeping his centre of gravity lower whilst tied to a more consistently lowered height of point of contact.

(Ranking, Pre: 9) Kei Nishikori

Utilising his base of support from the get go, this player not only flexes at his hips upon return, but manages to secure his balance and maintain a controlled centre of gravity. More prone to step up when returning with his backhand, whilst his forehand return tends to keep position, the player doesn't necessarily step up prior to contact, but isn't afraid from time to ever so slightly move angularly to cut off the angle. Exhibiting performance parameters of the elite *elites* through his return, higher consistency with his angular movement and meeting the ball inside the optimal zone of contact will see this return strengthen further.

The service motion of the player kicks into gear with solid lower body activation. Whilst upper body rotation is present, the player is limited in his hip flexion to initiate torque through his body. Further, as the player initiates his upwards drive, the upper body compacts to meet the height of the ball – a slightly higher ball toss to allow for a more pronounced upper body rotation to take advantage of his elastic energy would take this serve to the next level.

The player's groundstrokes typically resemble elite performance parameters with lower body activation present as with a consistently controlled state of dynamic balance achieved throughout play. Whilst the player has no qualm rotating through his shots on both sides, his ability to transition through the ball by setting his location of contact sightly further out in front remains inconsistent in its use, however distinct to the elite *elites*. For the player to recapture his height in the ranks, this smoothing through extension is key for further progression – although apparent at times, is yet to be consistently applied.

When considering the players movement, both horizontal and vertical positioning are integrated into the player's game. Angular movement is also applied, however limited in its application during play. Whilst shot anticipation and post-shot recovery appear to be fluid, the players use of momentum has room for further progressions – moving with the ball more often to take the offensive ground sooner opposed to staying where he is more evidently comfortable alongside the baseline.

(Ranking, Pre: 9) Kiki Bertens

Through return the player keeps a wide base of support whilst more often than not keeps her lower body activated through impact. This controlled state of the player's lower body works with the player's positioning to ensure she meets the ball at the optimal height of contact – near consistently on her service return. With an abbreviated take back at times – an optimal performance indicator, the player rotates through her return with ease, however does hold back on her momentum behind the ball which restricts her ability to extend on either her forehand or backhand through the ball. With limitations identified in the player's use of extension and momentum, both of these are tied to the player's proximity of the ball – catching the ball slightly earlier on her return would allow these limitations to be merely an end result of the player's progress in her anticipation of the ball.

When serving the player flexes her lower body to prepare to push her upwards to the ball, yet this flexion and reach is limited primarily by the player's limited flexion – also a result of her stance pre-contact. Although technically sound, the player tends to have a rather off-pace racket trajectory as she drops her racket face before lifting behind her head – from elbow through to a more smoother transition, yet this slight limitation in the player's fluidity puts her lower body off-pace and results in a near out of sync upper and lower body. Whilst a elite *elites* in other areas of her game, the player's serve, although consistent in its effectiveness, has its limitations for progressions. To have greater impact, the player is encouraged to slow down her racket take-back slightly and keep a more fluid motion to work with her lower body, whilst also increasing her range of movement at her knees and ankle to help the player further push upwards towards impact – contributing to her momentum behind the ball.

The player's groundstrokes maintain an optimally activated lower body paired with an optimal height of contact more often than not, identifying as a elite *elites*. With the player's optimal performance parameters through her groundstrokes, she is further able to rotate through her hips using torque to her advantage whilst she maintains a state of dynamic balance. Yet, the player does have a tendency to pull back off her shots and give up her offensive positioning. Further use of momentum, and heightened consistency in extending through her balls – taking her groundstrokes slightly earlier, would put this player in further contention inside the elite *elites*.

With horizontal movement favoured, the player does apply angular positioning from time through her groundstrokes, though this is yet to be consistent. Further, whilst the player does integrate vertical movement, this is dependent on the shorter ball as a result of the player limited in her natural use of momentum through her game. With further integration of momentum, paired with the player's use of angular movement, the player would more readily be able to extend through her balls, whilst also placing herself in a more offensive stance with her body behind the ball – in contrast to, at times, the player pulling off the ball and finishing early.

(Ranking, Pre: 10) Daria Kasatkina

The player has a sound serve, however is yet to optimise the use of her lower body. Although flexion is present, the upwards drive is near rushed and is in need of lengthening to establish the wind-up necessary to push the player upwards into the ball. This is compounded by a ball toss that is in need of moving further into the court to allow the player's body to drive upwards and ultimately fall into the court. Whilst rotation is present, the lack of generated momentum places unnecessary pressure on the player's upper body hindering her ability to fall into the court post-impact.

The player exhibits lower body activation through her groundstrokes and begins to exemplify a performance limited to the elite *elites*. However, inconsistencies creep through. More than capable of staying low to the ground with her centre of gravity well under control, the player loses sight at times of this need and raises herself – jeopardizing her optimal height of contact. This is perhaps the player's greatest weakness that then leads to her not transitioning through her shots and taking advantage of her momentum. Yet, the player is highly capable and whilst she exhibits optimal qualities her hips tend to stand tall opposed to flexing slightly to allow her to lean into the court, and through her shot. By further progressing her inconsistencies and beginning to lean into the court will see this player climb even high within the elite *elites*.

Upon return the player raises herself too early forcing her to make contact outside her most optimal contact zone. This limitation, a result of not moving up to, or away from the ball, places the player on the defensive to begin. Further, momentum into the court is scarcely present on the return, a direct consequence of not moving with the ball and meeting it inside the range of optimal height of contact.

A strength the player uses to her advantage is her depth of shot. This allows her to keep her opponent on the baseline and not easily move to the offensive. Further progress on her consistent use of her lower body in her groundstrokes, coupled with the use of momentum, will allow this player to have a more effective toolkit at her disposal.

(Ranking, Pre:11) Anastasia Sevastova

The player's return preparation exhibits slight lower body flexion, yet is predominantly hindered by a tall upper body with limited flexion at the hips. Whilst a wide base of support is present to begin upon returning, the player begins to lean back upon impact, preventing forwards momentum behind the return. Not shy of stepping up at times angularly to cut the angle off her opponents serve, the player manages to keep her height of point of contact controlled, indicative of an optimal performance of the elite. A slight forward lean into her return with flexion at the hip would see this player heighten her offensive positioning post return.

The player's serve is technically of the elite *elites* with deep lower body activation that propels her upwards and through impact of her serve. The player's elbow upon take back is limited in its drop which in turn limits her upper body rotation. A result of the player's base of support and her stored kinetic energy pushes her forwards into the court, using momentum to her advantage. A slight increase in the player's upper body rotation paired with her ball toss slightly more forwards into the court, would allow her to transition both her elastic and kinetic energy more effectively.

When contacting her groundstrokes the player manages to maintain a wide base of support with an activated lower body coupled with a predominantly controlled centre of gravity off both sides. A weakness becomes apparent with an inconsistent location of contact, although the ball height is predominantly controlled, an optimal performance indicator, her proximity to the ball at times is overstepped on her forehand that halts her from extending through the ball most optimally. Whilst the player opts for horizontal movement over angular on her groundstrokes, further integration of angular play would further optimise her groundstrokes and the player's all-round game effectiveness.

The player tends to move horizontally during play with limited vertical movement, a result of the player not using her momentum to her advantage in turn restricting her forwards movement. Although the player at times utilises her angular movement, this remains inconsistent. Whilst her proximity to the ball is predominantly consistent, anticipating the path of the oncoming ball to integrate angular movement, and momentum through her balls, would see this player move to the next level. This paired with a more purposeful flexion at the hips and placing her centre of gravity slightly more into the court, would then exhibit optimal performance parameters.

(Ranking, Pre: 12) Elise Mertens

When serving the player activates her lower body with good depth that allows her to push upwards to the ball. With the players upper body rotating through the ball, she has a tendency to loop at the drop of her elbow, in contrast to a down then up-through motion. At times this varied technique, although effective, catches the player's racket face more behind her back than most optimal contributing to the player extending over her head opposed to out and through inside her shoulder. Yet, ultimately the player exhibits optimal performance indicators, however, is slightly held back with her upwards transition through the ball and landing comfortably inside the court.

The player's return of serve keeps an activated lower body with slight flexion at the hips to keep her leaning into the ball. With the use of angular positioning through her return, the player isn't afraid to step forwards and cut the angle off her opponent's serve – yet, this isn't consistent. The player keeps her upper body in control as it guides the return with an abbreviated swing – an optimal performance indicator. Further, the player uses the kinetic energy in her lower body as her upper body works at controlling the ball. Often errors become commonplace when the player's timeliness is off, as with her use of angular positioning that also helps her finish through the ball. Yet, when the player's timeliness is in sync, her game more often than not identifies as it stands – a elite *elites*.

When contacting her groundstrokes the player exhibits optimal performance parameters with depth of flexion in her lower body, the presence of dynamic balance, as with an optimal height of point of contact. When pushed back, the player manages to keep her momentum forward to prevent her losing her offensive position. More often off her forehand, the player excels at opening her hips as with her use of angular positioning to not only cut the angle off the oncoming ball, but as to ensure her ease of rotation, before landing slightly forwards. Yet, the player's momentum behind her groundstrokes is held back, with a limited forwards transition, largely responsible from the torque achieved through her hips. With a backhand that also extends through, it does achieve momentum as a result of the player's torque with a slight bounce forwards – whilst the actual movement behind the ball is held back by the player more often. To take a solid hold on the elite *elites*, this player is in need of utilising her momentum to her advantage, whilst ensuring her depth in the lower body is maintained, with a tendency to lift at times which often is the cause of an error.

A mover that identifies as an elite, horizontal and angular movement are of ease, whilst the player does wait for the shorter ball prior to applying her vertical movement as she is most comfortable off the baseline. The player's anticipation of the oncoming ball is more often in alignment with her anticipation which puts the player at an advantage, scarcely found out of position. Yet, the player is limited in her momentum as she often holds back. Although the player identifies as a elite *elites*, with clear optimal performance indicators, and parameters throughout her game, slight progression in her moveability forwards, and open extension more consistently, will place this player in a more confronting position when competing against other elite *elites*.

(Ranking, Pre: 12) Borna Coric

Through his serve the player tends to stand slightly more front on that limits his ability to lean his hip into the court to optimise his torque to be released through impact. However, with pronounced lower body flexion that he optimises to transition into the court, using momentum to his advantage – an optimal performance indicator, the player makes up for this lean in his extended arch through his back that he uses to unwind and release through impact. Further, the player's extension is complementary to his ball toss as he is able to extend up and out through the ball where more often than not his momentum pushes him to an optimal post-impact position. The player does tend to over-extend his chest as he drops his elbow placing an increased load on his shoulder joint – this can be adjusted with a more vertical drop of the elbow in comparison that will work with the player to further pronounce his elastic energy through impact. Whilst the player near exhibits optimal performance parameters through his serve, as a near elite niche this is no surprise, yet for longevity at the ranks to stay on his side, as with a more consistent serve, slight tweaks in his extension and body alignment is key.

When returning the player widens his base of support prior to impact, yet his upper body is taller than most optimal limiting the player's ability to transition through and extend post-impact. A result of the player's stance inhibits his ability to gain the offensive through his return, predominantly caught on the defensive with his yet to be optimised return of serve. As a near elite *elites*, his return is his weakest part of his game that is yet to showcase his optimal performance parameters, consistently, as with other aspects of his game. Notably, when the player does anticipate the serve correctly, he does in fact execute the ball in alignment with his optimal performance parameters. Therefore, as a elite *elites*, consistency in his return is key whilst anticipation of the serve is a significant aspect to progress inside the elite niche.

The player tends to exhibit a lowered centre of gravity throughout his groundstrokes as with an optimal height of point of contact, although this is often susceptible to the player's positioning. From time the player's proximity to the ball is hindered which in turn results in the player applying limited flexion through impact whilst equally catching the ball higher than optimal and landing on his back foot. However, when the player's movement is in sync, so are primarily his groundstrokes as he then exhibits qualities of the elite *elites*. Evidently, the player's greatest weakness is in his consistency. Whilst the player uses momentum to his advantage and extends through both his forehand and backhand, at the same time as integrating angular positioning, he's restricted in his consistency, largely due to his proximity to the ball. To solidify his mark inside the elite niche, further awareness of his proximal development and ball location through impact, and its consistency, is key to allow his near optimised technique to deliver him the results warranted of his optimised technical parameters. Interestingly, when the player's proximal awareness is in sync with his proximity to the ball, he places himself in a position where he is near unstoppable – a consequence of his optimal performance parameters often apparent throughout his groundstrokes.

The player's current weakness is his consistency of play that is a direct consequence of his movement. Whilst an agile player readily able to move to the next ball, his movement exemplifies that of the elite *elites*. However, the player has a tendency to get caught through play being too close to the ball at impact which inhibits his ability to execute his desired groundstroke as most optimal. With horizontal movement of ease, the player's weaker aspect is his slight adjustments to or away from the ball vertically, affecting his ultimate proximity at impact. Yet, the player also has the advantage of his angular movement – an optimal performance parameter, yet remains inconsistent when the player's movement isn't in sync with the rest of his game. Perhaps most significantly, the player has phases where his movement, alongside his technical parameters align and he exemplifies the elite *elites*; whilst the player goes through phases where his proximity to the ball is out of sync that causes his game, to the contrary, to go out of sync as he attempts to scramble to find his feet. The player's integration of angular positioning is therefore key, as with his proximal awareness to the ball through play.

(Ranking, Pre: 13) Aryna Sabalenka

Upon return the player maintains a lowered centre of gravity that is then dropped slightly lower at contact – an optimal performance indicator. The player rotates through her upper body at contact, yet is limited in her ability to extend through her return at times catching the ball slightly late. At times the player does move through the ball on her returns, slightly, yet often pulls to the side post impact. Utilising the kinetic energy stored in her lower body with a slight upper body lean, courtesy of the flexion at her hips, sees the player more readily hit through her return with momentum – however, a result of the player's pull back results in an inconsistent return often placed on the defensive.

The player uses momentum to her advantage when serving courtesy of her ball toss and the upwards drive generated through her activated lower body. Whilst the player creates an arch in her back that is equally offset by the torque achieved, her lower body takes the pressure which is further pushed upwards to the ball. The player's upper body has sound rotation, whilst the speed of her release through contact is slightly limited. When extending through to the ball the player is limited at times with a slightly closer ball toss – not as inside the court as most optimal. Yet, whilst the player has optimal performance indicators through her serve, the consistently in their application is key, as with the need for a more pronounced flexion at the hips post contact.

When contacting her groundstrokes the player remains slightly flexed at her knees and at times meets her groundstrokes inside the optimal height of point of contact. Yet, the player's consistency of impact is countered at times by her dynamic balance not as steady – a result of the player catching the ball slightly inside, restricting the player's ability to extend through the ball. Whilst the player sets for the ball in a timely manner, the slight level of flexion in her lower body, although apparent, coupled with her limited flexion at the hips, keeps the player's momentum through the ball held back – limiting her ability to not only extend openly, but take a more offensive stance opposed to her predominant horizontal positioning. Although the player manages to rotate her upper body on her backhand and forehand, she does have a tendency to delay the drop of her racket face on take back off her forehand, whilst her backhand sees her positioned slightly closer to the ball. Off both sides, however, the player holds back – whereas momentum is limited, as with the extension through the ball. To progress, the player is in need of keeping her centre of gravity low consistently through transition, whilst integrating an extension through her groundstrokes opposed to pulling back and placing the pressure on her upper body.

The player's use of horizontal movement is most favoured maintaining her baseline stance. Yet, the player isn't shy of moving vertically when in need, although this is predominantly the result of her opponents' ball. A result of the player's limited use of her momentum sees the player restricted in her application of angular movement, opposed to cutting off the angle of the oncoming ball, the player opts to keep her horizontal positioning and maintain a defensive stance. To progress, the player is in need of integrating her use of angular movement to take an offensive stance more readily, coupled with her use of momentum whilst extending through her shots to move towards the technical parameters of the elite *elites*.

(Ranking, Pre: 14) Kyle Edmund

As the player returns serve he widens his base of support as he flexes at his hips in preparation to meet the ball out in front. At times the player slightly lifts through contact, although more often uses his kinetic energy through impact off both sides. With an abbreviated swing, the player takes advantage of his upper body rotation as he extends quite often through the ball. However, the player's ability to consistently transition forwards through impact is limited as he often pulls off. Yet, when he does manage to transition through his returns he exemplifies optimal performance parameters. The player's ability to extend through his returns is in itself an optimal performance indicator, highlighting the player's near elite niche calibre, currently limited by the player's level of consistency in his application.

When serving the player activates his lower body with an upwards lift as he extends to meet his ball toss inside the court. With a smooth technique, the player more often exhibits optimal performance parameters. However, the depth of flexion in the players legs is slightly limited at times restricting his ability to push upwards and out through his serve. The player rotates his upper body, contributing to his momentum achieved post impact – yet does appear to hold back on his release inside the court. With a slight increase in the player's range of flexion in his lower body, tied to his ball toss being slightly more inside the court at times, are key performance indicators to further optimize the player's serve.

Through his groundstrokes the player manages to keep an activated lower body through contact in sync with his more often than not optimal height of contact. The player often manages to rotate through his hips as he transitions through his shots, however is limited in his forwards momentum resulting in the player finishing with a slightly taller upper body. Yet, when the player intentionally leans into the ball and extends, he exhibits optimal performance parameters that quite often give him the offensive advantage. A more consistent application of the player's optimal performance parameters, in particular the player's ability to move with the ball, as with his use of angular positioning, are key for him to begin to take the offensive stance through play.

The player tends to maintain his stance on the baseline with horizontal movement readily on demand, whilst the player isn't shy of stepping up vertically to the ball, from time, although most frequently back-tracks to the baseline opposed to finishing off the point. The player does exhibit the use of angular movement from time, as with his use of angular positioning, although this remains an inconsistent part of his game. Further integration of angular positioning through the player's game will see his current optimal performance parameters move to the forefront, rather to their current conditional use. The player currently relies on his horizontal movement that limits his overall shot effectiveness as he tends to finish tall in position in contrast to a forward extension through the ball, courtesy of his at times use of angular movement.

(Ranking, Pre: 15) Ash Barty

Having had tremendous success in doubles, it has taken a number of years for that level of performance to transition to the singles court. This player has put in the work courtesy of her doubles conditioning and has come to the table with a sturdy lower body she is progressively using to her advantage. This becomes apparent during rallies on both sides, yet forwards and/or backwards movements are a weaker area – thus being at the mercy of the returners ball height opposed to having this movement in her arsenal for heightened control of play.

When looking at the return, here is a winner. Optimal lower body positioning is used in unison with the upper body whilst torque is applied. The player ensures the height of impact is in her hands whilst her movement works here – showcasing this is limited to her open rallies as it doesn't carry through to her returns. She is more than capable of this wider transition.

Behind the baseline the player's strengths begin to change. Although she can sporadically meet the ball within the appropriate zone, more often than not when pushed behind the baseline she disallows her height to be chosen. That is, opposed to moving forwards as earlier noted, the player remains and waits for the ball. This is a key flaw putting the player straight on the defensive regardless if control of the rally is in her hands, or not.

A strength at her disposal is something very few players have these days – a slice backhand that has the capacity to slide smoothly over the net opposed to a chop down motion. Although used in rallies at times from an undesirable height, tactics can't go amiss. The advantage here is in the player's take back and how the ball is met on its underside and the control of the strings against the ball. It isn't that it's overwhelmingly difficult, rather the player's ability to feel the ball on their strings. This is a strength of the player and one that will carry her into the elite rankings, further, when coupled with her vertical on court movements.

(Ranking, Pre: 15) Stefanos Tsitsipas

The service return of this player is near optimal – lower body activation is present as with his ability to stay down throughout the return and subsequent rally, utilising his stored elastic energy to his full potential. This is a key performance indicator. Further, the player uses angular positioning – not afraid of cutting the angle off the oncoming serve. Momentum is also on the players side, not shy of moving with the ball, and then recovering back to the baseline post impact.

Not afraid to move into the court with his groundstrokes, this player exhibits both horizontal and vertical movements through play whilst maintaining a constant state of balance. The control captured through his centre of gravity is that of the elite *elites* – although not quite there, yet. These are qualities at this stage of play that indicate optimal performance parameters. An area of slight weakness is the players location of contact – although he manages to maintain a mostly consistent height, at times the player is slightly too close to allow that smooth transition of the strings against the back of the ball, however, this is an aspect felt and not necessarily conditioned and will come with player progression.

The service motion of the player activates the lower body and drives upwards to the ball. Notably, his balance isn't as in sync during his service motion as with other parts of his game – falling wayside opposed to into the court at times. Upper body rotation is present, and solid; however, his ball toss does fall slightly behind of where it would be most optimally placed. Although slight, these are minor areas that are significant in this player's ultimate development to join the elite *elites* in the years to come.

Moveability around the court is a key performance indicator of this player, horizontally, vertically and angularly. This combination of movement parameters is often limited to the elite *elites*, or those on their way of progressing towards these ranks. A slight weakness becomes evident when the player is on the defensive, finding his way too far back behind the baseline to transition back into offensive play. Although widely accepted, the ability to transition between the two, defence to offense, is a key elite performance parameter to be further developed.

(Ranking, Pre: 16) Serena Williams

A player who has maintained her elite status for a solid two decades, the functionality of this player's all-round movement and technical parameters are exemplary. Her groundstrokes are maintained through optimal lower body positioning, and height, that allows her to reach through to the ball – not too close, for her strings to make impact behind the ball upon extension. A shortened take back allows the player to set herself early, taking time away from her opponents. Her solid base pivots through her shots whilst her constant state of dynamic balance is near mastered courtesy of her well controlled centre of gravity. A result of this style is a reduced forwards momentum, yet countered through her base of support, whilst her upper body takes the brunt of this load through its applied elastic energy – a rarity in elites yet a technical application that allows her to further benefit from her conditioned strengths.

When returning the player activates her lower body and transitions through the return – without lifting. This technique in itself affords the player to utilise her stored elastic energy to its fullest, whilst also maintaining a state of constant balance – not as secured when a player opts to lift. Further, her ability to still reach and hit through the ball is courtesy of her speed of reaction and anticipation of the serve. Ultimately, choice and awareness of height of point of contact becomes evident as the adjoining technical parameters fall into play.

A serve that is second to none, from lower body flexion, upwards drive, upper body rotation, to forwards momentum, complemented by a slowed down service motion – limited to the elite *elites* and even then not transparent. This player holds up until the racket is behind her head where her momentum takes over. Interestingly, most players including the elite *elites* can still overlook this slowed motion, whilst those outside this elite niche typically struggle to integrate and apply this sequence. A clean and simplified service motion, the player is capable of applying significant power through her serve courtesy of her overall technical parameters – achieving an optimal state of play.

Although weaknesses are scarce for this player, consistency is the biggest threat. In particular her lower body flexion and height of contact – at times compromised and her upper body takes over. This is tied to her timeliness of shot and being in sync all round – if her rhythm is compromised by her opponent, shot execution isn't as consistent or as effective. Yet, typically her opponents are unable to maintain their own level of consistency as to not let her find her feet. To the contrary, when her opponent does, there's a clear opening to take advantage of her height of contact – however, if inside her optimal height of contact, this player will make a meal of it.

(Ranking, Pre: 16) Daniil Medvedev

A smooth service motion that identifies as an optimal performance indicator as a whole, the player's lower body achieves a deep level of optimal flexion that is then utilised to drive him upwards to his point of impact. Yet, not before flexing at his hips does the player achieve a smooth rotation of his upper body. Kinetic and elastic energy are in play, as is the control of his centre of gravity that moves with the player in a controlled state as his service motion projects him inside the court. A slight progression would see the player lean his hip slightly more into the court to generate further torque that would then see the player generate further momentum to propel upwards and outwards into the court.

When returning serve the player achieves flexion at point of contact, yet catches his backhand slightly close to his body. Similarly, the player also catches his forehand return inside his reach unable to openly extended as a result of opting not to step up to the ball. Although commonplace, by waiting and not stepping up reduces the likeliness nor effectiveness of angular momentum and the player controlling his height of point of contact, consistently. The player sits slightly more upright as a result of limited flexion at the hips that results in a forward lean – placing his centre of gravity in a position still to be optimised.

The player's groundstrokes utilise his lower body achieving a solid bend upon contact on both sides, whilst the upper body rotation on his forehand, and backhand, is of the elite *elites*. Whilst the player exhibits optimal performance parameters, his forehand, although unique in a near-straight take back, it is rather the height of his elbow at times that compromises his shot outcome, meeting the ball above the optimal height of point of contact. Pairing this with a slightly more forwards lean through his shots, whilst taking into account his closeness to the ball at impact, are areas to address to progress further in the ranks.

Although the player holds a firm and well-rounded game, his proximity to the ball is a weakness. This is further countered by his anticipation and ultimate reaction to the ball in play. Although speed is on his side, it's the players location to the ball, slightly too close. This closeness limits the player to reach out and extend through the ball, rather getting caught slightly too close that limits his ability to transition through the shot, placing further pressure on his upper body to push through the ball opposed to utilizing his body as a whole. Although slight, this area of weakness once addressed will see this player move towards the elite *elites* with his current style of play.

(Ranking, Pre: 17) Madison Keys

When serving the player drops her lower body in rhythm with her upwards reach before extending through the ball. With an optimal service technique – fluid and simplified, the player achieves more often effective outcomes and places herself in an offensive position. The player's serve is near faultless as she uses her mechanics to her advantage, from the arch in her back offset by her lower body flexion, to the torque achieved that then rotates through upon contact. An area to consider is a greater range of motion in the player's lower body to work towards an even greater transfer of kinetic energy.

Upon return the player lowers her centre of gravity towards the ground to keep a sturdy base of support accompanied by flexion at her hips – leaning into her return. Not afraid to step up to the ball, the player often moves forwards prior to impact to expand on her transfer of kinetic energy through the ball coupled with her momentum. Exemplifying elite performance parameters throughout her return, the player extends through impact often limited to the elite *elites*. With good upper body rotation, the player anticipates the ball with ample time to swing through impact, whilst the consistent presence of dynamic balance contributes to the players consistent height of point of contact. At times the player occasionally catches the ball late that limits her ability to extend through the ball – an area for progression.

The player maintains a consistent state of dynamic balance throughout her groundstrokes as she often leans into her shots allowing her to transition through – often extending. At times the player is limited in her extension through the ball, however this is scarce and often when the player maintains a stationary position. More often than not the player is on the offensive with an optimal height of point of contact as with deep flexion through impact whilst she exhibits momentum through her shots. With optimal performance parameters evident through the player's groundstrokes, it becomes clear she possesses qualities of the elite *elites*, whereas a slight weakness at times is not hitting through the ball and finishing her swing too early.

A player who moves with ease all over the court, not only does she apply angular movement throughout her game, but the player is often moving forward as a result of her momentum behind her shots as well as her offensive stance. Further, with horizontal and vertical movement coming with ease, the player readily positions herself to ensure she can integrate her angular positioning to optimise the outcome of her balls. A slight weakness to consider for progression is the players lower body flexion as she is pushed back – tending to lift through her shots and her lower body, which more often than not puts her back in a defensive position.

(Ranking, Pre: 18) Garbine Muguruza

The lower body movement of her return of serve is second to none. Looking closer it is noticeable that although the body itself is switched on, the player holds back on her momentum into the return – her forwards movement. The choice of standing ground opposed to moving into the offensive in fact hinders the player in their rallies – a transitional flaw.

Predominantly height of contact is that of the elite, yet her horizontal movement is exceptional, the angular and vertical directional changes in fact contribute significantly to these downfalls, although slight. Upon closer analysis it is clear that when the player opts not to move angularly, rather sideways, her shots are restricted and more often than not the rally is lost.

The rotation of the player through her shots it commendable. Upper body torque is present on both sides, whilst her lower body is active the majority of the time when choosing her contact zone. This choice is most often a given "if" vertical directionals are not demanded.

When serving the player has a sound service motion with room for upwards push to further allow her to fall into the court – utilising her momentum. This is tied to the level of bend in her lower body – it is clear the player is holding back as she is more than capable of this in her groundstrokes. A factor of consideration here for both service and rallies is the location of the ball on contact – moving this contact point slightly in front will allow the player to utilise her natural strengths and progress deeper into the elite of the elite ranks.

(Ranking, Pre: 18) Milos Raonic

The player's return exhibits lower body activation of the elite *elites*. With evident hip flexion that results in a forward lean, momentum behind the ball is a natural outcome. The player manages go maintain a mostly optimal height of point of contact, more often on his forehand with his backhand at times allowing the ball to lift outside this zone. The upper body rotation achieved through the serve is a technique of the elite, turning from his hips whilst slightly abbreviating his backswing off both sides to ensure timeliness in his reaction and anticipation of the serve is present.

When serving the player exhibits optimal performance parameters of the elite *elites*. From lower body activation that incorporates flexion at the hips working to propel the player upwards to his ball toss, whilst incorporating an arch in his back to generate further torque. The technique applied by the player's strong upper body rotation sees his speed through impact achieved with ease. A slight limitation occurs from time when the player stops short of his full upwards lift that results in a more pronounced hip rotation post contact opposed to the necessary momentum into the court.

The groundstrokes of the player maintain a lowered centre of gravity with the consistent presence of dynamic balance. With optimal upper body rotation through his groundstrokes and the use of angular movement, the player is not short of taking time away from his opponents. Whilst utilising momentum to his advantage, the player moves through his shots further into the court more often than not. An offensive style of player, the player achieves near optimal proximity to the ball at impact, complemented by his consistent height achieved at impact – an optimal performance indicator. A limitation the player experiences is the ball falling off centre on his strings at contact, a consequence of the player from time overstepping his location of contact. This slight closeness to the approaching ball, although a limitation equally experienced by the elite *elites* when they typically exhibit optimal performance parameters, is a result of a misread in play anticipation, or in this case the player lifting his racket face slightly more vertically on take back that results in a slightly mistimed moment of impact with the strings slightly off centre.

The player's movement in itself is an elite performance indicator. The player eases through his horizontal movement courtesy of his consistently activated lower body and his controlled centre of gravity. Vertical movement is similarly applied with the player often making his way inside the court to finish the point. As with angular movement, the player typically moves angularly when stepping into the court with both his forehand and backhand benefiting from the use of angular momentum. Further integration of angular movement into the player's shots other than when moving to the net, would allow him to progress with his offensive play and take time away from his opponent.

(Ranking, Pre: 19) Grigor Dimitrov

A technically sound service motion that equates to the elite *elites*. Making the most of his lowered base of support, the player stores and releases his kinetic energy courtesy of his upwards drive that maintains momentum throughout his service motion. This is further attributed to his upper body rotation that draws on his elastic energy – overall, this player's technique exhibits optimal performance parameters, holding back slightly with his forwards momentum post-contact into the court.

The return of serve maintains a low base of support with a forwards lean indicative with his flexion at the hips. The player isn't afraid to move forwards, with the ball post-contact to transition through the shot, however this is yet to be a consistent response. Controlling his height of contact with slight movements, although the optimal height at contact when returning remains tangible. A current limitation is the player's angular movements when returning, currently scarce whereas a more consistent integration of this movement would allow the player heightened success.

When it comes to the player's groundstrokes these are also technically at their peak with an activated base of support, controlled height of point of contact courtesy of his movement towards, or slightly away from the ball, followed by a smooth rotation on both sides. Further, the players groundstrokes are complemented through his ability to set for the location of the oncoming ball that allows him to smoothly reach through and brush the back of the ball slightly before full extension, indicative of the elite *elites* performance indicators.

The player's movement is fluid with a controlled centre of gravity that delivers on the consistent use of dynamic balance throughout his game. The player has the ability to step up to the ball and directionally change its path as a result of his slight angular movements. Although not largely used, angular movement is applied from time, whilst horizontal is a clear strength as is his ability to move forwards into the court. Progress on his use of angular movement through his shots will push this player deeper into the rankings once again.

(Ranking, Pre: 19) Caroline Garcia

When returning the player keeps a low base of support coupled with an optimal height of point of contact, although does tend to lift from time that in turn limits her ability to transition through the return. The player rotates her upper body often with an abbreviated swing – an optimal performance indicator, as she also from time steps up to the ball and allows her momentum to transfer through impact. With evident optimal performance parameters, the player's central weakness is in her at times consistency in her height of point of contact, yet when the player does stay low she places herself into the offensive position more often than not. Not afraid to step up to the ball, when the player does, she often exhibits angular positioning and tends to be in control. With a slight improvement in the player's positional consistency, she'll continue to gain ground on the elite *elites*.

As the player serves she uses an abbreviated service technique which she relies on to transition her limited elastic energy through the ball. To compensate for her abbreviated swing the player achieves deep flexion through her legs which is also used to take the pressure off her back as she extends to create further torque through her serve. Whilst the player is limited in her elastic energy, she generates kinetic energy which is then transferred through her serve as a result of her upwards drive. The player is limited in her extension through the ball as she does have a tendency to hold back, also a result of her impact positioned at times closer to the player in contrast to slightly further into the court to encourage further extension and momentum post impact. To progress, the player is in need of moving her ball toss, as with turning her hips backwards slightly to allow her to let go and release through contact, allowing not only an increase in the players transfer of energy, but also her momentum behind the ball.

Through her groundstrokes the player frequently exhibits optimal performance parameters with a lowered centre of gravity tied to her optimal height of point of contact as she transitions through the ball. At times the player does have the tendency to lift through impact, or similarly, make contact slightly too high, limiting her effectiveness behind the ball. Yet, when the player does apply her optimal parameters, she exhibits qualities that once had her inside the elite niche. Capable of integrating angular positioning through play, the player is often on the offensive as a result of her key performance indicators. The player does have an unorthodox follow-through often on her forehand that does hinder her ability to extend through impact, however, when the player tends to keep a lowered centre of gravity, she does manage to extend through the ball finishing with her racket face around, opposed to above her head.

A player that moves with ease horizontally and vertically, her positioning is a key performance indicator of the elite *elites*. With the use of angular movement through play, the player varies her movement in relationship with the ball, often achieving near optimal proximity to the ball prior to contact as a result. Whilst the player comfortably moves forward, or backwards, to meet the ball inside the optimal height of point of contact, as with using her angular positioning to gain the advantage, at times she does catch the ball higher than optimal, or is pushed wide. To progress, further integration and consistency of angular positioning is key as is her steady state of a lowered centre of gravity.

(Ranking, Pre: 21) Anett Kontaveit

A sound serve with the necessary flexion and rotation, the player is further able to propel herself into the court post contact. Yet, there is depth to be further achieved in her lower body flexion to increase her upwards drive and its resulting forwards lifts. In addition, the player's use of rotation is similar, although applied, by working towards both her upper and lower body being in sync. By using her kinetic and elastic energy to her advantage, often seen as a limiting factor in pre-elite *elites*, will have a positive impact on the player's ultimate end result.

A return that keeps a steady base of support coupled with a balanced state, the player sets herself up to move with the oncoming ball. Yet, the player halts on her momentum transitioning back to the centre earlier than key that sees her lift slightly too early opposed to finishing through her shot and transitioning into an offensive position of play. This can be further progressed through an increase in the flexion at her hips upon return to keep her centre of gravity slightly out in front to balance the oncoming impact to be received by and through the server's ball. Further, this disposition puts her height of point of contact in a tangible position opposed to a controlled state whereby work on the former will progress this into a more actionable and effective state of play.

The groundstrokes of the player are smooth, whilst yet to be overwhelming. A relatively consistent style of play, the player struggles in her forwards momentum which at times then compromises her height of point of contact that in turns limits her overall shot outcome. More horizontally inclined whilst scarcely moving vertically into the court, the player's height of contact limits her ability to control play and transition out of the defensive. By progressing to activate her lower body and to maintain this flexion through her shots whilst working towards a slightly more extended contact, as currently her forehand in particular catches close to the body more often than not, will see the player further progress.

The player's level of anticipation and her ability to react is sound as she maintains her baseline position. Yet, by integrating a more purposeful use of angular movement would see the player move more often to the offensive opposed to her baseline comfort zone, and further progress her ability to react. Thus, by further developing and integrating her use of angular movement as with her ability to control her height of point of contact, will allow the player to deliver a more offensive and effective style of play.

(Ranking, Pre: 21) Nikoloz Basilashvili

Through his return the player keeps an activated lower body with slight hip flexion that allows him to lean into the serve. The player often steps up to the serve to use his momentum to counter the pace of the ball speed as he transitions through impact. From time the player does manage to extend through his return, although this is dependent on the player's anticipation. As the player rotates his upper body he uses an abbreviated swing off both sides that works with the player as he moves with the ball, although he is limited in his momentum behind the return. The player often exhibits optimal performance parameters through his return with a primary exception – angular movement. Not only will angular movement allow the player to step up to the ball and cut off the angle, but also work with the player to extend through impact.

When serving the player actively drops his lower body that is offset by a pronounced arch in his back as he extends towards the ball. A result of the player's lower drop and arch allows him to generate torque through his mid-section which is then released upon impact – an optimal performance indicator. The player rotates his upper body to align with his torque and its release through contact. Yet, the player does have a tendency to sway through his serve, often a result of his ball toss that results in his technical parameters and their effectiveness being compromised with at times a near off-balance stance post-impact. Further steadiness through impact, a consequence of his ball toss, as with a more open release through impact to use momentum to his advantage, are key for the player to advance his service technique.

The player maintains a consistent presence of dynamic balance through his groundstrokes as he actively drops his centre of gravity prior to, and post impact off both sides. The player tends to extend through his forehand from time, encouraged through the evident release through his hips, whereas his backhand is slightly closed off that inhibits the player from openly extending post impact. With an optimal height of point of contact often achieved through his forehand, favoured by the player over his backhand, from time the player's proximity to the ball is overstepped which causes him to lift through impact, whilst also inhibiting the player's ability to extend through impact. However, when the player sets, not only does he integrate angular positioning, he also manages to extend and release his torque through impact.

To advance, consistency is key for the player as with his ability to gauge his proximity to the oncoming ball. Further, from time the player will make impact above the optimal height of point of contact and lift through his legs – more timely positioning to anticipate the ball as with his ability to move to, or away from, to achieve his optimal height of point of contact, will allow the player to take the offensive position more readily.

With horizontal movement frequented by the player as he sticks primarily behind the baseline, he adjusts to the oncoming ball quite often by moving vertically backwards, opposed to taking the offensive stance forward, as he then transitions through the ball. From time the player integrates angular movement through his forehand, primarily, as his backhand, although sturdy, is often run around when speed is on his side. A more prominent use of angular positioning, as with his ability to move forwards through the ball – using momentum, is key for this player to progress. Likewise, sticking closer to the baseline is where the player exhibits optimal performance parameters, yet the further he moves back, the higher his point of contact becomes, as with the flexion in his lower body.

(Ranking, Pre: 22) Jelena Ostapenko

When returning the player maintains a low centre of gravity with a predominantly controlled height of point of contact – an optimal performance indicator. The player isn't afraid to step up to the return to re-work her positioning to ensure her height of point of contact remains most optimal – a facet of the elite *elites*. Further, the player rotates her upper body on both sides, yet also abbreviates her take-back as needed to ensure her timeliness of impact meets the ball out in front, opposed to applying a full open swing resulting in the player catching the ball too close to her stance. Thus, the player's awareness of the proximity of the ball at impact is indicative of an optimal performance parameter, whereas to progress, further use of the player's momentum behind the ball, as with a more pronounced drop in her centre of gravity, will see the player move deeper within the elite *elites*.

The player exhibits an activated lower body through her serve, however is limited in her upwards drive to the ball, resulting in the player landing post-impact near behind the baseline, at times, or slightly out in front. However, a result of the player's use of her upper body rotation and speed through transition, sees the player benefit at impact from her release of elastic energy. Whilst technically sound, the player is slightly limited in the flexion at her hips, which is further compounded by her lower body flexion – although present is limited in its depth which contributes to the player's restricted release of kinetic energy to push her upwards to impact. To progress, the player is in need of a more pronounced bend in her lower body, whilst moving her ball toss slightly inside the court, contributing to the natural release of the player's momentum behind the ball.

The player exhibits a consistently activated lower body through her groundstrokes with flexion present through her forehand and backhand – pre and post impact. With a controlled height of point of contact readily achieved by the player's movement in her preparation for the ball – whether slightly backwards, forwards, or a variation of angular positioning, the player exhibits optimal performance parameters. The player at times tends to fall off her shots post impact, whereas although the player extends through her shots, opposed to moving with the ball has a tendency at times to pull back. A limitation is evident in the player's ability to apply momentum behind her groundstrokes – although readily able to apply angular movement, the player often opts to restrict her natural pull into the court which when held back, halts her ability to readily take a more pronounced offensive position.

With the capacity to move horizontally, vertically and angularly as required, the player exhibits optimal performance parameters more often than not, a direct consequence of her all-court moveability. Whilst the player is held back at times, a result of her transition through her shots, more often she is readily able to extend as a result of her angular positioning that contributes to the player's more often offensive play. The player readily has the ability to cut the angle off the oncoming ball, as with her ability to take time away from her opponent – courtesy of her early positioning. Whilst optimal performance indicators are more often present, the player is at times restricted in the execution of her discrete skills, which ultimately affect the serial skill in play, for example, the player's use of momentum behind her groundstrokes, as with her serve.

(Ranking, Pre: 23) Carla Suarez Navarro

When returning serve the player slightly activates her lower body coupled with a lean into the court, yet interestingly the player begins to slightly drop her centre of gravity further, prior to contact, with a near optimal height of point of contact achieved. However, upon contact the player does in fact begin to lift, resulting in limited lower body flexion that further restricts the player's use of her kinetic energy through the return. The player exhibits good upper body rotation with an abbreviated return – an optimal performance indicator, although is limited in her forwards momentum. To progress, the player is in need of integrating angular positioning – often catching the return too late which keeps her in the defensive position. Further, by beginning to use angular positioning, the player will naturally be encouraged to transition through the ball that will provide offensive positioning.

The player's serve achieves optimal lower body flexion that is used to drive the player upwards to the ball – an optimal performance indicator. Yet, her lift up to the ball is limited with her ball toss being closer to the player's body, restricting an open release of her kinetic and elastic energy. The player does achieve good rotation in her upper body, as with an elbow drop that then extends upwards and outwards towards impact. Whilst technically sound, generating a more pronounced drive upwards with a ball toss slightly more inside the court, the player would not only benefit from her body's natural momentum, but further release her kinetic energy into the ball.

Upon contacting her groundstrokes, the player manages to stay low to the ground on her backhand – as a single handed player, she is able to keep a steady presence of dynamic balance whilst her lowered centre of gravity sees the player lean into the ball to transition through. However, the player is limited post contact in her movement, opting to recover sideways opposed to through the ball openly. Likewise, the players forehand keeps a low base of support – an optimal performance indicator, as with a controlled height of point of contact, whilst her transition through the ball is again limited. Whilst the player closely exhibits optimal performance parameters through her groundstrokes, she is often held back as a result of her momentum coupled with her limited extension through the ball – a result of maintaining her pre-contact positioning opposed to moving with the ball. To progress, movement behind the ball to take a more offensive stance would see the player move out of her often defensive positioning to gain ground and step up to her opponents.

The player is a baseline player who tends to be most comfortable off her groundstrokes. With a preference for horizontal movement, and at times vertical – although often backwards as a result of controlling her height of point of contact; whilst exhibiting optimal performance parameters technically, the players weakness is clear in her positioning – keeping back and quite frequently maintaining a defensive position. The integration of angular movement throughout her game would enable the player to not only move to a more offensive position, but cut time away from her players. To progress, further emphasis on the player's anticipation of the ball would encourage her to adapt her movement and see her climb higher in the ranks.

(Ranking, Pre: 24) Roberto Bautista Agut

Upon return the player keeps an activated lower body, yet has the tendency to lift through impact, dispersing his stored kinetic energy. A consequence of the player's near upwards stance through impact places a heavy burden on his upper body as it absorbs the pressure. However, the player does manage to slightly transition through his shots which allows him slight momentum behind the ball. With an abbreviated swing on his return – an optimal performance indicator, the player does slightly flex at his hips allowing him to lean into the ball. From time, the player does manage to stay low to the ground through impact and achieve an optimal height of point of contact – yet his consistency wavers. A greater level of consistency through his return as he aligns with his optimal performance parameters will provide greater return success.

When serving the player activates his lower body before extending upwards towards his ball, exhibiting smooth technical parameters through his serve. A result of the player's ball toss primarily located inside the court allows the player heightened momentum behind the ball as he flexes at his hip through impact, using his generated momentum to pull him inside the court. Yet, the player is limited with his extension through impact, as with the release of his elastic energy, a consequence of his often present flexion at the elbow through impact in contrast to a clean open release through the ball. The player does exhibit optimal performance indicators through his serve consistently, rather is limited in their resultant impact due to his limitation in his transfer upwards and through the ball. Progress on this build and subsequent release will work with the player to generate a more effective service outcome.

The player's groundstrokes primarily exhibits a lowered centre of gravity, however the player varies between a controlled height of point of contact, higher than most optimal that sees him lift through impact, to groundstrokes off both sides that do in fact align with optimal performance parameters for both lower body activation and height of point of contact. The player's weakness is in his consistency of application and further, an underlying reason why the player tends to be a defensive player – limited in his ability to transition through the ball and extend to move up closer to the baseline and take an offensive position of play. With the tendency to stay well behind the baseline, the player's strength is in his speed of movement and racket control. A strength of the player is his ability to push his body as a whole behind the ball as he maintains a compromised stance through play, working with his weakness and the significant 'why' the player is kept on the defensive. A more purposeful integration of an activated lower body through play as with the player extending through his groundstrokes will allow him to progress to achieving a more offensive game to consistently deliver against the elite niche.

With movement as his strength, the player manages to run down a significant number of his balls, courtesy of his agility. Yet, with horizontal movement favoured as with his deep baseline positioning, the player scarcely uses angular movement as he tends to stay further back. Interestingly, when the player opts to step up closer to the baseline, he not only further activates his lower body and achieves an optimal height of point of contact, he begins to take a more offensive stance that often places his opponents on the back foot. Therefore, the player is in need of stepping up to the baseline more consistently, as with his use of angular positioning, to allow his optimal performance parameters to shine through.

(Ranking, Pre: 25) Dominika Cibulkova

When serving the player lowers her centre of gravity tied to a slight arch in her back before lifting towards the ball, whereas her ball toss is positioned inside the court to generate optimal momentum behind the ball – an optimal performance indicator. The player rotates her upper body in sync with her lower body flexion, whilst as she begins to drive through her legs, her elbow begins to extend – exemplifying a technically optimal service motion. However, the player's weakness becomes apparent upon closer analysis of her lower body flexion – at times this doesn't begin until the ball is at its peak before coming through. The lag in this discrete skill has a roll-on effect to the serial skill – the serve, and at times the player's ability to drive upwards through the ball.

The player activates her lower body as she prepares to make impact with her groundstrokes, yet does raise herself prior to contact with the ball. With a mostly controlled height of point of contact, the location of contact is optimal when timed in alignment with her positioning – out in front to allow her to extend through the ball, more often off her forehand. The player exhibits a consistent presence of dynamic balance which is called on when pushed behind the baseline as she lands backwards, yet reacts forwards into the court to regain ground. Yet, her height of point of contact is slightly above most optimal at times, whist she does inconsistently make contact most optimally when she has ample time on her side. With good upper body rotation off both sides through the ball, the players greatest limit is in her deep lower body flexion – rarely present, coupled with a consistently controlled height of point of contact. With limited angular positioning, a more pronounced use of this movement as with the players anticipation of the ball are key initiators to enable her to progress to a more consistent use of these optimal performance parameters.

Upon return the player primarily achieves an optimal height of point of contact as she impacts the ball, coupled with an activated lower body. Using an abbreviated swing through her return – an optimal performance indicator, the player rotates through her upper body, although does from time catch the ball too close to her body, limiting her ability to extend through the return. Positionally, the player tends to maintain her position and not move forwards through her return, or angularly, limiting her ability to take an offensive stance. Further depth in her legs by a lowered centre of gravity, as with the integration of angular positioning are key indicators to further progress the player's extension through her return.

An all court mover that will move vertically when presented with the shorter ball, yet predominantly does stay on the baseline, agility is a tool the player isn't shy of using. Yet, as horizontal movement is favoured, the player has a tendency to find herself in the defensive position behind the baseline. A result of her agility, however, enables the player to quickly recover to the centre, yet she is limited in her offensive stance more often as a result of angular positioning being scarce. To progress, angular positioning is key as with her ability to maintain flexion at her hips to lean into the court and extend through her shots.

(Ranking, Pre: 26) Camila Georgi

The player exhibits solid upper body rotation through her serve whilst her lower body begins slightly activated before deeper flexion is present in sync with the player's elbow drop, prior to swinging up to the ball. Although a slight delay in the player's upwards lift is apparent, she does manage to use her kinetic energy upon her drive to the ball. The player manages to generate momentum into the court, largely a result of her ball toss and the rotation of her upper body. Yet, at times the player appears to hold back from reaching up to the ball and subsequently flowing through her serve to benefit from the use of momentum and kinetic energy available, apparent from the delay and depth of her lower body flexion. An increase in the use of the player's lower body would contribute to the use of torque in her serve which would ultimately contribute to the release of her elastic energy behind her serve.

The player begins her return of serve lowered whilst as the ball approaches she drops slightly further, lowering her centre of gravity to achieve a near optimal state of balance upon return. The player's lowered state at contact, alongside her positioning, contributes to her optimal height of contact when returning. However, at times the player's proximity to the ball limits her ability to extend through the return and step into the court, opposed to pulling off sideways. Further, her consistency is held back by her willingness at times to move up to the ball, whilst although exhibiting a return close to the elite *elites*, incorporating upper body rotation, this optimal performance indicator is limited as a result of the players closeness to the ball.

When contacting her groundstrokes the player keeps her centre of gravity low to the ground, however her consistency is at times limited. Keeping her lower body activated on both her forehand and backhand, the player manages to transition through her shot's courtesy of her use of vertical movement, although inconsistent that sees her use of momentum held back. A result of the player at times holding back leaves an opening for her opponents to take an offensive stance, yet when she opts to transition through her shots the player exhibits optimal performance parameters – consistency is key. With upper body rotation that allows the player to hit through her shots, as with the presence of dynamic balance, optimal performance parameters are evident with this player. Yet, the player's weakness is in her consistency of application with a large contributing factor being her use of momentum, opposed to remaining near stationary at impact, paired with at times her closeness to the ball which hinders her ability to extend with a release into to the court.

Not afraid to move into the court and keep her offensive advantage, the player moves vertically with ease behind her balls when moving into the court. Favouring horizontal movement with a preference to stay on the baseline, glimpses of angular movement come through, although scarce as the result of the player's limited use of momentum through her groundstrokes. The player's proximity to the ball is well controlled, although a weakness comes through in her controlled height of point of contact raising at times as a result of the player favouring her horizontal positioning. The positioning of the player is adaptive as she transitions to a more offensive stance, whilst to progress the player is to heighten her level of consistency when hitting through the ball, as with her height of contact, to head towards the elite *elites*.

(Ranking, Pre: 27) Lesia Tsurenko

Upon return the player has slight lower body activation at times that sees her put on the back foot limiting her transition through the ball. When the player activates her lower body with depth, at times, she manages to stay low through the ball and is more effective in her execution. The player utilises her upper body as she transitions through the ball with solid rotation at impact. Although her momentum is scarce as a result of limited flexion at the hips, the player's lean into the ball allows her base of support to maintain her control. Whilst limited in her momentum and largely lower body flexion, the player's height of point of contact is sound, yet coupled with moving up to the ball would progress the player further.

The player has a sound service technique with limited lower body flexion that halts her upwards drive. Whilst upper body rotation is present, the player's elbow remains slightly raised on the drop back, coupled with the position of her ball toss closer to her body than most optimal. The player's ability to effectively use her drive is restricted which results in limited momentum post impact. An increase in lower body flexion would position the player to further drive upwards extending her lift, whilst also offsetting the arch the player generates in her back. Pairing this with the player's ball toss slightly outwards, would begin to then exhibit optimal performance parameters.

When it comes to the player's groundstrokes her lower body does manage to remain activated throughout her shot. The player manages to rotate her upper body on both her backhand and forehand sides, whilst slight hip flexion is present when the height of point of contact is controlled. Typically moving horizontally along the baseline, the player is limited in her vertical movement that prevents her from stepping up into the court and towards the oncoming ball to control her height of contact. This inconsistent height of contact with her infrequent vertical or angular movement are current limitations holding her back.

Whilst the player manages to move effectively horizontally along the baseline, vertical movement and its resultant momentum is scarce. The player is restricted in her shot anticipation as she remains near stationary, horizontally, for the next ball – this is opposed to stepping inside the court and meeting the ball early at the optimal height of contact, or for the player to slightly move backwards, before moving forwards again, to ensure that optimal height of contact is achieved. The integration of angular movement into the player's game, as well as the progression of extending through her shots would begin to exhibit optimal performance parameters in her game.

(Ranking, Pre: 27) Denis Shapovalov

As the player serves he achieves deep flexion through his lower body, leading with his hip as he leans into the court – an optimal performance indicator. As the player extends up towards the ball his elbow drop is slightly restricted as he tends to over-extend through his chest that places a higher load on his shoulder as a result of the player's slight chest-extension. Yet, the player does manage to use momentum though his serve, as he also drives through his lower body up towards the ball. To progress and to achieve a higher consistency in his serve, a slight adjustment to the players torque is key. By increasing the use of his midsection, the player will rotate back as he extends, before releasing into the court, whilst keeping his upper body slightly less extended at times to take the load off his shoulder joint. Further, a slightly more inside position of the player's ball toss will allow for a more technically optimal technique. Whilst the player does exhibit optimal performance parameters primarily, it is this slight adjustment as a result of the load on his shoulder and over-extension that will offer the player's serve greater consistency, and the player's body greater longevity.

When returning serve the player exhibits slight knee flexion before he begins to use his momentum through the ball, lowering his centre of gravity even further towards the ground – an optimal performance indicator. With hip flexion present as the player goes to return, he transitions through the ball whilst often extending more frequently on his single-handed backhand with his forehand, at times, limited in the player's momentum behind the ball. Interestingly, the player integrates angular positioning as he returns, from time – an optimal performance indicator as he readily gauges his position to the ball. Yet, the player's anticipation of the return can be caught out from time, as with his rackets proximity to the ball – not starting his racket low enough to then lift up and around through the ball resulting in error. Further, whilst the player often exhibits optimal performance parameters, these are often dependent on his position – if the player's proximity to the ball is slightly misjudged, he does begin to lift through impact and often creates the error.

The player maintains a consistent presence of dynamic balance through his groundstrokes as with a widened base of support indicative of deep knee flexion present through impact, as with a predominantly controlled height of point of contact off both sides. With the tendency from time to lift through his forehand and exhibit slight flexion, as with an upper body pull-back, the player has room here to build a more consistent technique under pressure. The player often transitions through his shots, however does pull-back post-impact on his forehand from time as a consequence of his proximity to the ball. Yet, on his backhand the player openly extends as he uses momentum more often to his advantage and maintains his depth of flexion in contrast to his forehand from time. To progress, the player needs to use his momentum post-impact more consistently whilst maintaining his flexion at the hips through his forehand more often as with his widened base of support.

The player often exhibits a variety of movement through play in particular his use of angular movement, although not yet cemented firmly in his game. With horizontal movement coming with ease, as with his adjustments vertically, slightly forwards, or backwards, the player's lowered base of support keeps his centre of gravity balanced throughout play. Yet, when the player opts to position himself angularly his game is that much more effective – not only allowing the player to step up to the ball and extend through openly, but release his torque through impact as he rotates through his hips and takes time away from his opponent. A primary area to advance is the player's proximity to the ball that results in him lifting through impact – slight adjustments and heightened awareness of his levers and racket face with the location of the ball will readily fix this inconsistency as with a constant lowered base through impact particularly in these weaker instances.

(Ranking, Pre: 28) Fernando Verdasco

When serving the player lowers his centre of gravity with pronounced knee flexion that offsets the arch in his back used to further generate torque through his midsection – an optimal performance indicator. Further, as the player reaches to extend towards the ball, his ball toss is primarily inside the court that allows the player to use his momentum through impact. With a near technically optimal serve, the player does from time keep a ball toss slightly closer to his body that results in a land on the baseline post-impact – indicative of his limited extension through the ball, as with his momentum. Although the player rotates smoothly through his upper body, he is somewhat restricted which, as with his level of hip flexion post-impact, halts back his serve from being one of the elite *elites*.

Through his return the player exhibits slight flexion in his lower body with limited flexion at his hips that results in a taller upper body than most optimal. A result of the player's stance is his limitation in transitioning through the return and using his stored kinetic energy to counter the pace of the ball – a consequence of his slight flexion. Whilst the player does tend to use an abbreviated swing as he rotates his upper body, a consequence of his higher point of contact at times is his limited capacity to extend through his return and apply momentum resulting in a defensive return often leaving the court open for his opponent. A slight move towards the baseline and through the return, as with his ability to maintain a lowered base of support, would progress this player's return to be more effective against the elite *elites*.

The player exhibits deep knee flexion through both his forehand and backhand with an optimal height of point of contact, although his consistency in application varies. With a tendency to lift through impact at times that results in his kinetic energy transferring upwards, rather than through the ball, as with keeping slight knee flexion tied to a higher point of contact – often the case when the player is behind the baseline. Yet, when the player does step up to the baseline and dictate play, he exhibits optimal performance parameters – consistency for this player is key. Similarly, his proximity to the ball from times can be cause for the player lifting too soon that sees his upper body take the load at impact. Whilst the player transitions through impact, his extension and post-contact momentum is scarce. Favourably are the player's technical parameters that when applied more often than not achieve a more desired outcome.

Predominantly a baseline player, horizontal movement is primary called on by the player whilst his vertical movement is used when adjusting to the oncoming ball, more often backwards behind the baseline. At times the player does show his ability to integrate angular movement, as with his rotation and subsequent release through his hips. However, a result of the player's primarily horizontal movement is his consistent defensive position that is reliant on his opponent to make the error in contrast to stepping up to the offensive and forcing the error from his opponent.

(Ranking, Pre: 29) Maria Sharapova

When serving the player activates her lower body in sync with her upper body extension followed by her lift towards the ball. With good upper body rotation that is compounded by her ball toss primarily being inside the court, the player is able to transfer her stored kinetic energy through the ball. An apparent weakness at times is the player's ball toss location as she goes to reach towards the ball. Yet, the player does exhibit optimal performance parameters for her serve as she transitions through the ball. A more prominent lower body bend including flexion at the hips to push the player upwards and through the ball, as with a slightly lower ball toss, at times, will allow this player's serve to reach new heights.

Upon return of serve the player leans into the court and slightly extends behind the ball. With an activated lower body and controlled height of point of contact, the player exhibits optimal performance parameters through her serve returns. Not afraid to use her angular positioning, the player from time will rotate her hips in line with the ball to allow her lower and upper body to push through the ball as one – an optimal performance indicator. Whilst the player excels at her returns, at times she does have a tendency to lift post contact that sees her balls go long, or at times, into the net – a variation in shot outcome to do with the player's timing of the ball on her strings. Further development on the player's location of contact and timing will bring more consistency to the player's return outcome.

The player's groundstrokes maintain a lowered centre of gravity as with a controlled height of point of contact – more often than not. Exhibiting optimal performance indicators through both her forehand and backhand, the player has a tendency to lift slightly before contact to not stay as low which in turn sees a higher point of contact. Although this happens from time, it is an inconsistent part of her game. The player does use her hips through her groundstrokes with torque generated, as with momentum behind the ball – an optimal performance parameter. As an elite the player's greatest weakness is her inconsistency at times, yet equally, her ability to perform optimal performance parameters are second to none when in sync and consistency is there. To progress, it is key for the player to maintain a lowered centre of gravity that in turn will see her keep her height of point of contact low – this tied to her transition through the ball will see her reach the pinnacle of the elites once again.

The player often integrates angular positioning through play – an optimal performance indicator. Whilst primarily preferring to maintain her baseline positioning and thus horizontal movement, the player isn't afraid to move vertically as she frequently integrates angular movement throughout her respective points. As the player primarily keeps a lowered centre of gravity as with flexion at the hips, this allows the player to move forwards through her balls and extend – a natural body transition with this stance and further indicative of an optimal performance indicator. At times the player is slightly inconsistent with her angular positioning which allows her opponents to hit past the player – further anticipation of the ball, as with the player more consistently stepping out to rotate her hips to set before impact, will not only see her angular positioning take the advantage, but further optimise her already acquired optimal performance parameters.

(Ranking, Pre: 30) Aliaksander Sasnovich

When returning serve the player activates her lower body and controls her centre of gravity with a slight forward lean, although limited. The player's height of contact upon return is higher than optimal that sees her stored kinetic energy go upwards opposed to through the ball. A slight progression in moving upwards to the ball whilst maintaining a wide base of support would see the player control her height of impact and reduce the pressure currently placed on her upper body.

The player's groundstrokes see her lower body remain activated throughout her shot transitions with the capacity to use her momentum through her shots as a weapon. Not easily pushed back the player moves both horizontally and vertically through play. Her height of point of contact is perhaps her most significant strength courtesy of a well controlled centre of gravity that allows the player to maintain her wide base whilst transitioning through her shots, exhibiting the role of dynamic balance. The player's forehand has a wider take back that can see her time to contact reduced, yet she utilises the elastic energy gained to her advantage when her timing is in sync. It is key for this player that her movement is in sync as with her proximity to the ball to ensure these restrictions have a limited impact on her overall shot outcome.

The serve of the player identifies with a sound technique, yet the positioning of her ball toss, at times, slightly limits her end result coupled with the limited flexion between her lower body and hips. This slight progression in technical arrangement halts the player from achieving the necessary upwards drive to take her serve to the next level. Whilst the player is able to achieve upper body rotation and momentum inside the court, the limited cohesion between the knees and hips to propel her upwards and out is merely holding the serve back from increasing in its delivery of the player's stored kinetic energy, further.

When anticipating play the player is comfortable moving horizontally along the baseline, yet is limited in her want to move forwards to take a more offensive stance. Whilst the player is more than capable with her vertical movement, she does tend to fall back on contact at times which keeps her in a defensive state of play. A slight move of her centre of gravity to encourage a forward lean accompanied by momentum into the court to transition through her shots are performance parameters in need for the player to further progress.

(Ranking, Pre: 30) Gilles Simon

Upon return the player activates his lower body and rotates through his upper body – although with an abbreviated swing identifying with an optimal performance indicator. However, at impact the player does have a tendency to stand taller and limit the flexion present in his lower body, restricting his ability to transition his kinetic energy through the oncoming ball. At times the player does integrate angular positioning, although isn't a consistent part of his returns. With the player at times applying optimal performance parameters throughout his return, his greatest limitation is in his consistency of return. The use of momentum behind his returns is scarce, whilst further consistency in his application of a controlled height of point of contact and having his lower body remain activated at and through contact, are primary performance indicators to further progress.

The player achieves good upper body rotation through his groundstrokes with a predominantly controlled height of point of contact off both sides as with a lowered centre of gravity through impact. Yet, the player's ability to extend through both his backhand and forehand is scarce, primarily wrapping around his groundstrokes without a forwards transition. Whilst the player rotates his upper body, it is often closed off and limited in its ability to openly transition through his shots. Likewise, is his use of angular momentum through play, typically positioning himself on the baseline. To progress and to take a more offensive stance during play, momentum through the player's shots is key which is instigated by angular positioning that then rolls on to the player more so extending through his shots – key for the player to near the elite *elites*.

When serving the player achieves slight lower body activation as he rotates his upper body, however opens his hips up to the court slightly too early placing a heavier reliance on his upper body as it goes to impact through the ball. Whilst the player reaches towards the ball he arches back which is only slightly supported by his lower body – a result of its limited flexion. Further, with a ball toss not as optimally placed inside the court, this places increased pressure on his upper body. Although the player exhibits a smooth and fluid service motion, as he restricts his upper body from opening up, tied to his already slightly forwards hips, the player has a more front on service motion at times which limits his ability to openly transition through his serve and utilise his kinetic and elastic energy to his full advantage.

As a baseline player, his movements are predominantly horizontal unless the short ball approaches – only then will the player move vertically into the court. However, whilst the player is strong off both sides, his limitation in his forwards momentum holds him back from his angular positioning. Therefore, from angular positioning to extending through his shots and the player's momentum behind the ball are all key for further advancements and are instigated by the player lowering his centre of gravity with a more consistently applied flexion at the hips to pull the player into the court and position him behind his shots – opposed to maintaining his neutral position.

(Ranking, Pre: 31) Alex De Minaur

A service motion that activates the lower body and achieves upper body rotation, the player holds a smooth serve with a purposeful upper body rotation. His ability to achieve momentum through his service motion is heightened through his upwards drive and use of his kinetic energy propelled into the court post-contact. It is clear, however, that the player's lower body flexion is still limited as with his use of hip rotation to achieve sturdier torque through his shot.

When returning the player stays low to the ground with a wide base of support whilst maintaining an inwards lean – elite performance parameters. The height of impact is also controlled with the player not shy of moving ever so slightly forward followed by a bounced-back recovery. Hitting through his return is achieved, yet at times closer than ideal – anticipation and readiness have work to do as does the location of his point of impact slightly further in front to allow that extension to hit all the way through the ball.

The groundstrokes of this player, although yet to reach the elite *elites* ranks, hones in on their optimal performance indicators. The player's ability to keep a sturdy base of support throughout the rally, whilst his centre of gravity is kept increasingly controlled courtesy of his low base of support throughout, permits the player to consistently utilise his kinetic energy at near all times, whilst his elastic energy is consistently released through the rotation of his upper body through his shot.

Limiting factors that the player faces is not with his movement – readily able to react and move on the court with speed on his side. Yet, as a defensive player his ability to move up into the court isn't his strength which at times halts the player in his use of angular movement, delimiting his shot outcomes. Further progress on his lower body activation through his serve, ball proximity on the return, and increased use of angular movement in his groundstrokes to take time away from his opponents, will see this player join the elite *elites* in no time.

(Ranking, Pre: 31) Katerina Siniakova

Upon her return of serve the player is limited in her lower body activation, making contact with the ball with merely slight flexion in her lower body that is then further lifted as she ends with a taller than optimal stance post contact. However, from time when the player anticipates the serve more readily, she does manage to keep an activated lower body and hit through the ball, although is inconsistent in this application. Whilst the player does rotate her upper body through the return, she is often caught with a slightly late point of contact. Yet, in the player's favour is her optimal height of point of contact more often than not, whilst although an optimal performance indicator, her lateness in impacting the ball, although slight, and limited lower body flexion at impact, holds the player back from matching the elite *elites*.

When serving the player manages to lower her centre of gravity, although slight, before lifting towards the ball. The player creates an arch in her back that is limited by her lower body activation to counter the load, resulting in the player's upper body rotation taking further pressure. Further, as the player goes to drop her elbow prior to extending to the ball, she opens up her racket face and allows her wrist to drop slightly back – not only does this slightly offset the player's fluidity and control of the ball, it further takes away from the player's ability to smoothly transition upwards and outwards into the ball. Although technically the player sways in her fluidity, she is able to gain momentum behind her serve courtesy of her ball toss often being inside the court. To progress, the player is in need of addressing specific technical parameters around her serve to achieve a smoother motion, coupled with further activation of her lower body.

The player isn't afraid to step forward on her groundstrokes, with her lower body activated pre and post contact coupled with an optimal height of point of contact. Whilst the player applies optimal performance parameters through her groundstrokes, her anticipation of the returning ball has a tendency to catch the player too close to the ball. However, although the player's proximity to the ball at times is a limitation, this is compounded by her fall-back often on her backhand, a result of the player catching the ball too close to her body. The player does exhibit optimal performance indicators from time off both sides, however her consistency in their application is the player's primary limitation. Whilst the player rotates through her groundstrokes, her extension is limited – typically a result of the player's slight backwards-counter that holds the player back from a forwards transition.

The player's court movement varies in its use – primarily a horizontal mover, the player isn't shy of stepping forward when the opportunity strikes. However, a result of her often defensive position sees the player on the baseline, or behind, most often – this is furthered by the player's backwards-counter post impact. Opposed to falling through her shots, the player falls back – her body's way of ensuring she maintains a state of balance. Yet, as the player is limited in the forwards lean at her hips, the result of action-reaction – her body counter responds backwards. A result of the player's at times backwards movement restricts her from not only moving through the ball, but her chance to apply angular movement through play. Therefore, by integrating a lower base of support to promote flexion at the hips, coupled with an integration of angular movement – stepping up to the ball, are primary factors that are key for this player to progress in the rankings.

(Ranking, Pre: 32) Lucas Pouille

The player's ability to move into and up to the ball is perhaps one of his key performance indicators. When returning the serve, he isn't shy of moving up to the oncoming ball – a key elite parameter.

In good company is his service action – upper body rotation is present, lower body uplift, and forwards propulsion. It's an elite service motion. Slightly more questionable is the player's recovery from their service endpoint, at times the returner placing the ball deeper than the server's actual court positioning. Therefore, the bounce back isn't quite as strong. Fortunate for the player, too often returners neglect to note this and merely becomes troublesome against the elite of the elite.

The player's defensive position behind the baseline does the job – whilst his height of contact isn't desirable, he manages to counter this with his upper body strength. Not desirable for longevity, however achieves short term results.

Movement isn't shy with vertical directional changes not a problem. This level of ability, in particular speed forward, is a key positional advantage and suggests angle cut offs aren't an issue. All round a player with this level of movement has a greater capacity to do more damage in the rankings long term.

(Ranking, Pre: 34) Donna Vekic

Upon return the player establishes lower body flexion, slightly, yet interestingly lowers herself further upon impact. Whilst the player at times moves up to the ball for impact, whilst also controlling her height of point of contact, the momentum behind the ball and transition through her return remains slightly restricted. Yet, the player's near consistent height of impact identifies as an optimal performance parameter with the flexion of her lower body paired with her optimal height of contact. This is further enhanced by the player's solid upper body rotation and shot anticipation allowing the player to hit through her returns without restraint.

The player has a smooth service motion with a sound technique, yet the degree of activation in her lower body limits the player from identifying at this time as a elite *elites*. Yet, the player does counter this limitation in lower body depth with near optimal use of her elastic energy. Whilst the player exhibits torque through her service motion, interestingly the player uses an abbreviated service motion – limited in rotation and typical elbow flexion. Yet, the location of the player's ball toss coupled with her use of elastic energy achieves momentum for the player allowing her to not only finish inside the court, but with effective hip flexion to transition and extend through her service motion.

Clear optimal performance indicators are apparent during the player's groundstrokes with optimal lower body activation paired with a near consistent height of point of contact. The player's ability to move towards the ball and adjust her stance slightly whilst preparing for impact, or slightly backwards whilst countering her movement with the presence of dynamic balance, exhibits optimal performance parameters. The player's use of her upper body and its open rotation further contributes to her forehand and backhand performances that allows the player to gain a more consistent offensive position. A limitation at times is clear in the player's extension through the ball where she opts to flex her wrist inwards at times opposed to extending through to the end of her reach.

The player's level of anticipation to the oncoming ball is heightened by her level of agility applied in adjusting to the ball prior to impact. Both horizontal and vertical movements are of ease, whilst angular movement is also frequently applied when the player initiates vertical movement, whilst not as commonplace when the player opts to maintain her baseline positioning. Further use of momentum through the player's shots would see the player's upper body open up, whilst full extension would begin to be of ease. Likewise, by further use of her lower body in the serve would generate an upwards drive to take advantage of the player's kinetic energy.

(Ranking, Pre: 35) Nicholas Kyrgios

The service motion of this player achieves a fluid and rhythmic motion. From his lower body flexing, hips turning to achieve torque, to his upper body rotation, the player uses his streamlined motion to let go upon impact – unleashing his stored kinetic energy. The player's greatest downfall here is his limitation in each respective area – room for a deeper lower body drive, an ever so slight backward twist with his torque, allowing his upper body to increase in its range of motion, before hitting further through the ball. Although a clean and near elite serve, the player does hold back with room for a deeper technical arrangement and slightly lengthened service motion.

Utilising his elastic energy throughout his groundstrokes, the ball achieves solid rotation on his forehand, yet is limited on his backhand side. Despite his efforts to activate his lower body, this is only slight that results in a higher point of contact and less controlled stance. The player's centre of gravity keeps the player in a balanced state, however is limited in his forwards lean resulting in a more off-balanced state of play when confronted with deeper balls. A consequence of his lower body not fully activating is the player's location of contact, not only predominantly higher than most optimal, but his ability to reach for the ball and subsequently transition through – limited as the player often rushes his shot as a result of not being in the most optimal position pre-contact.

When returning the player has a tendency to sit up with limited activation of his lower body, or lean into the ball – hip activation is limited, forcing his upper body to take the brunt at impact. Although there are glimmers of stepping into the court pre-contact, the player more often lifts prior to impact, limiting his forwards transitions. It becomes noticeable upon returning the player opts to move back to the centre of the court post-contact 'before' completing his actual intended return – this rushed behaviour limits the player's overall controllability of his return of serve.

The player utilises horizontal, vertical and angular movement to his advantage. Capable of applying all positions and cutting off the angle of the oncoming ball, his at times spatial awareness to the ball is his only slight limitation – stopping him from hitting through the ball at times by over calculating his set position. A consequence of not consistently activating his lower body is his ability to take off and speed of movement – by not staying closer to the ground this slightly hinders his overall moveability. Although slight, by activating his lower body more consistently and with intention, would begin to close the gaps in his game, overall.

(Ranking, Pre: 35) Daria Gavrilova

When returning the player manages to activate her lower body, however her centre of gravity remains slightly high at contact. The player's height of point of contact is slightly higher than most optimal, a result of a tall upper body at contact with limited presence of hip flexion that would result in the player having a more pronounced lean into the return. Whilst upper body rotation is present, the player at times catches her return slightly late which further halts her ability to extend and transfer through the ball. Further, when approaching the return, the player's anticipation of the ball is inconsistent whereas her proximity to the ball is slightly overstepped which causes the player to catch the ball on her strings at times, further contributing to limiting the player's extension through the ball – more often making contact is followed by her body weight pulling to the centre, opposed to behind her return.

The player manages slight lower body flexion on her serve which inhibits her transition through the ball at impact. Whilst the player rotates through her upper body and extends upwards through contact, she is restricted in her upwards lift more often than not. The player's limited use of her lower body inhibits optimal transfer of kinetic energy, yet although her upper body transfers her elastic energy, the inconsistency in her use of momentum behind the ball holds her serve back, tied to her slightly open strings at impact. With the presence of an arch through her serve, the player's limited lower body activation prevents the build of kinetic energy with her back taking the load at times, placing a heavier reliance on her upper body.

The player manages to time her groundstrokes to set prior to impact, yet this work is at times hindered by the player pulling up on the ball opposed to extending through, largely on her backhand. When contacting her forehand the player has a more agile movement in preparing for the ball, yet as with her backhand, is limited in her extension tending to wrap around post contact with restricted use of momentum which also tends to keep the player on the back foot in a defensive position. Whilst the player has a relatively smooth technique, she is hindered by her transfer of energy through both sides of her groundstrokes and her tendency to keep an upright body which further inhibits the player from benefiting from her kinetic energy, let alone the use of momentum through her shots. Similarly, the player's proximity to the ball needs to be further moved in front to promote extension through the ball which will also force the player to begin to lean from the hips rather than keeping her current taller upper body stance.

The player's movement is relatively sound, whilst although at times she exhibits comfort in her vertical movement when moving to the offensive, this is yet to be a consistently formed aspect of the players game. Predominantly, the player maintains her baseline positioning which is further compounded by her limitation in moving through the ball and her use of momentum, keeping her near idle alongside the baseline. Therefore, whilst horizontal movement is favoured, the player does illustrate her use of vertical and at times angular movement, yet remains inconsistent as a result of the player's primary limitation in moving with and through her groundstrokes. To progress, the player is in need of reworking her technical parameters post-contact whilst integrating a more activated lower body through contact coupled with her ability to extend through the ball openly.

(Ranking, Pre: 36) Marton Fucsovics

The player initiates his serve with flexion present in his lower body, however his speed of transition is halted by his held-back drive through the ball, slowing his speed through contact. With a ball toss positioned closer to the body than most optimal, the player limits his momentum into the court with limited flexion at his hips. Yet, the player manages to control his serve with a relatively sound technique. The arch apparent in the player's back is somewhat compounded through his lower body, yet as the level of depth is limited, this contributes to the player's limitation in his upwards lift to the ball. A more activated lower body with a ball toss slightly positioned inside the court, to begin, will contribute to the player's serve progressing to a elite *elites*.

Upon return the player does exhibit slight flexion in his lower body, however this is limited. The player's anticipation of the serve is at times restricted that results in a delayed reaction to the ball coupled with a higher than optimal height of point of contact. Whilst the player rotates his upper body through the return, heightened pressure is placed on his upper body to push through the ball as a result of his lower body remaining limited in its activation through impact. Further, as the player opts to maintain his pre-return positioning more often than not, his momentum behind the ball is restricted as he tends to pull off to the side as he recovers to the centre of the court prior to transitioning through his return. To progress, the player is in need of a more lowered centre of gravity coupled with the use of angular movement to position himself more optimally as he works to extend through his returns.

When contacting his groundstrokes the player manages to keep slight flexion present in his lower body, yet his upper body remains quite restrained as with his shoulders that halts the player's ability to extend through both his forehand and backhand. Whilst the player does exhibit the ability at times to transition through his shots, there is an inconsistency present here that limits the player's consistency in performance. Momentum is applied from time behind his groundstrokes, yet is also limited as a result of the player maintaining his baseline positioning more often than not in contrast to the integration of vertical and angular movements through play. At times the player's proximity to the ball limits his ability to extend – his closeness inhibits the player's ability to take a more offensive stance. To progress, integration of angular play as with a more pronounced lower body through contact, paired with his ability to extend through his shots, are primary performance indicators to heighten his ranking.

The player's court movement is largely horizontal with scarce use of either vertical or angular movements. Whilst the player has a sound technique, he is limited more often in his effectiveness through play as he holds back on both his extension through the ball and momentum. Primarily responsible for the player's limitations is his court positioning and responsiveness to the oncoming ball. Progression in the player's anticipation of the ball, as with a lowered centre of gravity to ease through a more active court positioning, are initial performance indicators to consider as with his ability to extend through the ball, to contribute to a climb in the rankings.

(Ranking, Pre: 37) Andreas Seppi

Solid lower body flexion is present through the player's serve paired with a smooth arch in his back, countered by his evident knee bend. Yet, as the player goes to extend towards the ball, his upwards drive is limited, more often relying on his upper body and the release of his elastic energy, to drive through the ball. Whilst the player does exhibit momentum through his serve, the limited flexion in his lower body holds the player back as his upwards drive is slight, resulting in the player relying on the extension of his upper body to release through his serve. It becomes apparent that whilst the player is limited in his application of optimal performance parameters, he does exhibit a somewhat smooth service technique that is primarily held back by an initial closed base of support that contributes to the players limited lower body activation.

When contacting his groundstrokes the player exhibits a lowered centre of gravity, however begins to rise at contact. Although the player's height of point of contact identifies as an optimal performance indicator, his rise at times through his lower body halts the player's transfer through the ball. Solid off his backhand, the player exhibits strong lower body activation paired with his upper body rotation, yet his momentum is scarce. Similarly, with the player's forehand, although he manages to activate his lower body, his ability to extend through the ball is limited. To progress, the player is in need of extending through his groundstrokes which will see the player begin to use his momentum behind his balls. Further, with a slightly more activated lower body and integration of angular movement, the player will begin to transition to the elite *elites*.

Upon return of serve the player activates his lower body, yet his upper body remains relatively tall. Whilst flexion at the hips is present, as with the player's upper body rotation, the player tends to be more closed through his return that limits his ability to transition through the ball. At times the player will move to step up to the return, however this is scarce, primarily keeping back whilst limited in his extension, tending to pull off and set for the returning ball. To progress, the player is in need of a more open body through his return as with a slightly more pronounced lower body activation. Bringing the player's centre of gravity slightly down will encourage the player to naturally move with the ball, as he begins to integrate more often, optimal performance indicators.

With a preference for horizontal movement, the player is limited in his use of vertical and angular positioning. Although managing to keep a lowered centre of gravity consistently which contributes to the presence of dynamic balance through his game, the limitations remain in his extension through the ball as with his use of momentum. The integration of angular movement would allow the player to look at optimising his positioning to move towards an offensive stance in contrast to the player's primary defensive position through play. Further, the player's preparedness can look to further be optimised with the player's anticipation of the ball, slightly held back as a result of the player's limited use of angular positioning when setting for the oncoming ball.

(Ranking, Pre: 37) Johanna Konta

When serving the player steps up to her serve as she flexes her lower body before extending up and through the ball. Often the player is caught extending with a bend in her elbow in contrast to a full extension to openly release the player's elastic energy through the ball. Whilst the player does achieve momentum post impact, the location of the player's ball toss can hinder the player's ability to achieve full extension and come down on the ball – at times over rotating her upper body opposed to extending through. Although slight, progression of the player's ability to extend through impact will place her serve in a more offensive position.

As the player returns she keeps a low base of support with evident flexion at the knees and hips – contributing towards her forwards lean into and through the ball. With an abbreviated swing quite often on her return, the player also transitions through – although is limited in her extension. With an often controlled height of point of contact, from time the player does lift through impact that often sees her balls land beyond the baseline. The player uses her upper body to control the ball as she rotates and releases her elastic energy through the ball – when she keeps her lowered stance; a consequence of the player lifting through impact is the displacement of her kinetic and elastic energy and their transfer through her return.

The player primarily exhibits optimal performance parameters through her groundstrokes with the presence of dynamic balance as with a controlled height of point of contact compounded by the depth of flexion in her lower body. A weakness is evident as the player from time will rise through impact as her height of contact and base of support begin to rise, loosening her control on the ball. Yet, more often than not the player will transition through her groundstrokes. When it comes to extending, the player will from time, when dictating play; whereas when the player is in a rally she does tend to wrap around her groundstrokes opposed to hitting through and openly extending.

An all-court mover, the player doesn't hesitate as she moves forwards vertically, or horizontally along the baseline. From time the player does integrate angular movement as she sets for her attacking balls, whilst primarily through her rallies she is limited in her angular movement, scarcely stepping up on the angle to cut the angle off the ball. Yet, as the player does from time integrate angular movement through play – an optimal performance parameter, she does exhibit qualities of the elite niche, merely limited in her consistency of application of these set performance parameters. To progress, the player is in need of a higher level of consistency of integrating angular play, whilst also looking towards a more streamlined use of her extension through the ball in both her serve and groundstrokes.

(Ranking, Pre: 38) John Millman

The serve of this player sees a heightened arch in his back upon contact that is not entirely offset by his lower body, although flexed, limited in its upwards drive. Likewise, the rotation present in his upper body gets the job done, however is limited in its effectiveness without a more significant transition through his lower body. Whilst the lower body is active, the upwards drive is slowed due to the applied arch coupled with the player's ball toss slightly back restricting his momentum into the court. A result of this ever so slight limitation in a fluid motion is the player's reduced use of his kinetic and elastic energy and their contributions to the service outcome.

When returning serve the player manages to move up to the ball, however his base of support begins to lengthen. This reduced level of depth in his lower body halts the player's ability to move through and with the ball, typically placing pressure on his upper body to return the ball. Further, this tallness upon impact limits the player from achieving an optimal height of contact consistently, affecting his ability to control the balls location and extending smoothly through his return, opposed to the increased pressure placed on the player's upper body that is limiting the impact of his returns.

The player's groundstrokes utilise a solid base of support that allows the player to control his centre of gravity consistently. With a preference to move horizontally along the baseline opposed to forwards and/or in an angular position, this limits the player's effectiveness in transitioning between defensive and offensive play. The player's forehand sees his elbow rise slightly more than most optimal that takes time away from the player's preparedness as well as limiting the player's ability to benefit from the role of torque as he remains limited in the transition of his hips through impact.

A player that possesses enduring on court speed, his limitation largely remains in anticipating the next shot and the resultant angular movement – opposed to remaining on the baseline with his favoured horizontal movement patterns. Therefore, speed nor agility is on this player's side, yet his ability to move angularly is a weakness that keeps this player in a predominant defensive state of play. Progress on the application and integration of angular movement, whilst remaining low through his shots more consistently, are key performance indicators required to see this player move his ranking towards the elite.

(Ranking, Pre: 39) Frances Tiafoe

The return of serve is restrained in this player's case, largely due to his limitation in moving away from the oncoming ball – getting caught either in line with the ball, or not moving away fast enough to ensure he has adequate reach, but isn't too close to the ball to transition through the ball at impact. This is tied to his lower body – although activated at impact, he isn't transitioning through the ball as most optimally, mostly due to his positional stance in relationship to the ball. The player's positioning upon return tied to his limited use of his lower body, restricts him from altering his stance and having a more effective and diversified return.

The player's serve sees a slight arch in his back that isn't adequately supported by his lower body bend – this is problematic long term. This choice of service motion can be offset with an increasingly activated lower body to ensure the player's base of support takes the load away from his back – in turn driving the player upwards and into the ball. As for the player's upper body, it is equally sound with present rotation, yet without the drive of the lower body there is plenty of room for progressions.

When it comes to his groundstrokes, technically he's in a very solid position. The player does in this instance use his lower body most optimally and rotates through his shots, however gets caught positionally if the ball doesn't come back to his hitting zone. Rather, the player's movement is typically horizontal, although has captured the use of angular positioning. Both positional stances are lacking anticipation – the player is yet to find his footing when it comes to moving to, or away from the ball, with speed, likewise moving up to the ball to cut off the angle of the oncoming ball.

The player's agility is there, as for his court speed – his control over his centre of gravity allows him to use his shots to his advantage. Yet, the limitations of his serve, return and groundstroke spatial awareness require attention for the player to have a greater impact on his opponents and a climb in the rankings.

(Ranking, Pre: 39) Venus Williams

When serving the player activates her lower body coupled by and upwards extension towards the ball. However, the player is slightly limited at times on the flexion in her lower body that limits her upwards drive towards the ball. Although the player exhibits clear optimal performance parameters through her serve, her forwards momentum is limited as she holds back on her drive towards the ball, indicative of her slight position inside the court post-impact or at times landing on the baseline. A slight increase in flexion of the player's lower body as with her forwards extension through the ball will bring this player's serve back to where it once was when inside the elite niche.

Through her return the player tends to stand rather tall with limited flexion in her lower body coupled with a higher than optimal height of point of contact. Yet, on both her forehand and backhand, when the player opts to step up to the ball she does in fact exhibit optimal performance parameters in height of point of contact and a pronounced flexion in her lower body. Therefore, the player's consistency of application is key. Further, the player does tend to fall off her return as she recovers to the centre opposed to extending through the return. With the player's upper body in control when set, the player's anticipation of her return is a clear weapon – yet, dependent on the player's positioning. A more consistent level of play through her returns to ensure she's applying her optimal performance parameters will see the player's return more often near unstoppable.

The player primarily exhibits an activated lower body through both her forehand and backhand with a near optimal height of point of contact. The player does tend to contact her balls at times higher than most optimal which often results in the player extending through her lower body prior to impact. Yet, when the player does stay down and maintain her optimal performance parameters, more often than not winners are the end result. The player's current weakness is in her consistency of contact, as with her ability to consistently transition through her shots and extend through the ball – often holding back on her momentum behind the ball. Whilst at times the player does exhibit qualities of the elite *elites*, where she once reigned, her tendency to hold back, as with her consistency of her technical parameters, currently halt the player from progressing inside the elite niche once again.

An all-court mover, whilst the player is strong off the baseline and in her horizontal movement, equally the player is in her vertical movement, in particular as she often moves up to attack a shorter ball, or to finish the point off at the net. The player's movement identifies with the optimal performance parameters of the elite *elites* with her speed of transition and anticipation of the oncoming ball. Yet, the player is also currently limited in her use of angular movement, as she tends to play through the middle of the court, or horizontally – without cutting off the angle of the oncoming ball, or setting her stance to face inside the court.

(Ranking, Pre: 40) Jeremy Chardy

The player maintains a solid base of support when returning serve with slight flexion at the hips, whilst still sitting slightly taller than most optimal. Predominantly waiting for the serve to come to him, vertical movement post-return is scarce as with forwards momentum. The ability to move angularly and cut the angle off his return is present, however its effectiveness is slightly held back as a result of his taller stance – whereby moving his centre of gravity slightly forward to increase his forwards lean would further push the player behind the ball and benefit from his body's natural momentum, placing the player in a more offensive position post-return.

A service motion that utilises the step-up approach works in this player's favour with a pronounced level of lower body flexion that works to push him into the court. The upper body rotation coupled with his use of flexion at the hips upon reaching exhibits performance parameters of the elite *elites*. The location of the ball on his strings at contact can at times drop slightly low, as with his racket face maintains a slightly open face – this combination sees his balls hit a fraction open that results in his serve landing short in the court. Further, this slight technical flaw inhibits the player from utilising his kinetic energy to its fullest potential.

Looking at the player's groundstrokes, he is well versed in the use of his lower body and relatively consistent with an optimal height of point of contact. Whilst exhibiting optimal performance parameters technically, the player is limited in his anticipation of his opponents' ball and his readiness – more often than not positioned well behind the baseline reducing the time the player has to react to a shorter or more angular ball. Yet, the player will at times move angularly and vertically, however transitioning between baseline play to a more offensive position is scarce as he relies on his opponent to draw him forward.

The player has an all-round game, whilst his weakness is in his want to move to the ball, consistently. With glimpses of offensive play that see the player dominate points, this is dependent on his willingness to step up to the ball, and his timeliness in ensuring the location of the ball is within his reach, opposed to being too close. A near elite *elites*, the player is limited in his application of angular movement and contact through the ball, dependent on his positioning. Progress on the player's anticipation and all-round timeliness will allow him to build a more offensive style of play.

(Ranking, Pre: 41) Maria Sakkari

A solid service motion with leg drive, upwards lift and forwards momentum. The player is technically sound and utilises her mechanics to her advantage. A slight move of the ball toss would allow the player to increase her upwards drive in a more forwards lean opposed to the slight arch she opts to create that sees her momentum push more vertically opposed to integrating a more angular lean. Further, her racket face opts to remain open throughout (slice), more often, where a slightly inwards turn with the ball toss closer to her centre, would see her edge towards a more optimal service motion.

Staying low more often than not on her groundstrokes, her 'loopier' forehand backswing, with a lifted elbow, hinders her shot outcome and ultimately her timeliness to make contact at the optimal height of impact. Her backhand is streamlined, however, forwards momentum on either side isn't a consistency. Further, at times the player lifts during her shot, more often her forehand, an area for progression to stay down to prevent an uncontrolled outcome.

When it comes to her movement, both horizontal and vertical are both present as is angular positioning. The player overall exhibits a sturdy centre of gravity and well controlled state of balance through play. A downside is her ability to transition between defensive and offensive play. Although highly capable, her anticipation of shot and choice of placement is an area keeping her from the elite *elites*.

The player's return is solid, not being shy to step up and into her return. Utilising her lower body to hit through the ball is an optimal performance trait, however her height of contact varies at times that places her slightly on the back foot. Momentum is present as is her ability to recover. Yet, throughout the player has a tendency to slightly lift post-contact – by working towards a more consistent lowered base will be advantageous to her strengths and afford the player a more offensive game.

(Ranking, Pre: 41) Martin Klizan

Standing slightly tall on his serve return, the player is limited in his lower body activation yet manages to achieve slight flexion upon contact. With hip flexion present with a forward lean, positionally the player is in a sound position. The player's height of point of contact is relatively controlled with the use of angular movement, yet this still remains an inconsistent part of the player's game. Further consistency on the player's height of contact coupled with his use of angular movement will place him in a better position to also progress his location of contact with his proximity to the ball at contact slightly close to the body, restricting the player from extending through his returns.

A service motion that has the player's hips closed to begin, the player does manage to open out upon impact with a heavy reliance on torque and upper body rotation. Although achieved and a technique that works for the player, a result of this initial position is a slight disadvantage to the player's overall momentum and transfer of kinetic energy. Whilst lower body flexion is present, the range is limited and yet to be as pronounced as most optimal. Upon contact the player achieves upper body rotation, yet the elbow remains slightly bent throughout impact limiting the players transition through the ball and ultimate momentum into the court.

The player maintains an activated lower body throughout his groundstrokes exhibiting the use of dynamic balance. With the ability to hit through his groundstrokes, they are limited in their transfer of energy – a consequence of the player remaining quite stationary post impact. The player tends to raise his elbow on the forehand side higher than most optimal which takes slightly more time away from the player when preparing for the ball. When in position the player has the ability to keep an optimal height of point of contact, although this range isn't consistent throughout, either falling back on his forehand at times, or remaining 'stuck' post contact on his backhand limiting his transfer of energy or use of momentum to maintain an offensive stance.

The player uses both vertical and horizontal movement quite well, yet is limited in their post-contact use and the roll of momentum. Whilst horizontal is favoured, the player isn't shy of moving into the court, yet is restricted in his movement 'without' the consistent integration of angular movement through his shots, opposed to placing added pressure on his hips and rotating through and transitioning with the ball. The player is predominantly limited in the effectiveness of his groundstrokes by holding back – not pushing into the court. A result of the limited momentum applied through his shots sees the player keep a near defensive stance, whilst his location of contact is too close, restricting his ability to extend through the ball, whereas currently the player near wraps-around the ball, further compounding his limitation to move through the ball.

(Ranking, Pre: 42) Adrian Mannarino

Upon return the player exhibits slight lower body activation, whereas hip flexion is present pre-contact, the player contacts with a taller upper body that results in a centre of gravity closer to his body. This closeness of the player's stance results in a limited transfer of energy and has the capacity to compromise the player's state of balance as with his ability to return the ball, opposed to extending through. The player manages to exhibit an optimal return take-back at times with an abbreviated swing with a smooth transition courtesy of his upper body rotation. The player's height of point of contact at times is controlled, however, is outside the optimal contact range too, with consistency in need for progressions. Further, as a result of the player's taller upper body at contact and limited lower body flexion at times, the player is limited in his transition through the shot, as with his momentum post impact.

The player's serve has near optimal lower body activation that drives him upwards to the ball. Whilst the player exhibits torque in the climb towards his serve, as with upper body rotation, he remains quite tall above the hips with limited hip flexion. Post contact the player does exhibit the use of momentum as he lands inside the court, however opposed to an inwards 'dive' the player lands with limited knee flexion then bounces back – indicating limited applied momentum. The player's upper body achieves good rotation, whilst his chest remains quite closed opposed to opening up in the player's upper body extension. This tendency to stay closed, with his slightly closed hips at contact, limits the player's initial upwards lift to the ball before keeping his momentum restricted, further contributed by his reach into the court.

The player contacts his groundstroke with knee flexion present in his lower body, whilst his contact zone is within the ideal range, although yet to reach a consistent optimal height of the elite *elites*. The player's groundstrokes exhibit an activated lower body pre and post contact, an optimal performance indicator, with a controlled height of point of contact, yet his transfer of momentum is limited. Likewise, although the player applies his lower body and benefits from his stored kinetic energy, the player is limited in positioning his centre of gravity further out in front to transition through the ball, rather the player remains quite stationary post impact and is reliant at times on the push of his upper body. The proximity of the player to the ball currently limits his ability to hit through the shot, as he tends to wrap-around his groundstroke, equally off both sides, without extending through. Further, as the player opts to move horizontally more often with limited integration of angular movement, this halts the player from a more natural use of his momentum as with his ability to lean into his groundstrokes, further contributing to the presence of dynamic balance through his shots.

A steady baseline player that can comfortably move horizontally, the player is limited in his use of vertical movement largely as a result of his limited momentum through his shots. Whilst there are scarce instances of angular movement, this is yet to be a consistent part of the player's game. The location of contact for the player is compromised as at times he is pushed back from the court as a result of his limited use of momentum, keeping him on the defensive. To integrate a more offensive stance the player is in need of integrating momentum into his game as with his anticipation of the ball to allow the player to react accordingly and move more angularly opposed to remaining at times stationary on the baseline.

(Ranking, Pre: 44) Yulia Putintseva

Sitting up high for the return and waiting for the serve to come to her results in the player being forced to contact the return well above most optimal. Yet, on the off chance the ball comes to her with her already positioned stance, her lower body is activated. The player is held back, however, on her return by not moving into the ball, putting her on the defensive from the get go.

Solid upper body rotation is apparent in her service motion, however her lower body remains to be activated, although ever so slight. This limitation in her lower body prevents the player from lifting off into her ball toss and releasing that stored energy (power) into the ball. A direct consequence of this gap in the lower and upper body is a service motion that is overly arched, unnecessarily, and whilst sound, leaves the player on the back foot with her opponents able to take full advantage of the reduced ball speed coming their way.

The player's clear weapon becomes apparent in her groundstrokes keeping low to the ground with a consistent activated lower body. This is furthered by her height of contact and location – although positionally could be slightly more out in front to achieve that racket head glide over the ball upon impact. Her horizontal movement is solid, yet there's room for her to move up to the ball opposed to sitting on the baseline waiting.

Anticipation and angular movement are in need of further development. Whilst the player is solid off her baseline, her weakness is in the anticipation of balls and the timeliness of her reaction. This is tied to her angular movement – scarcely applied. For progression into the ranks, these two performance indicators are key alongside her lower body activation largely in her service motion.

(Ranking, Pre: 45) Alize Cornet

An all-round player and a tour veteran, her strength is in her consistency and ability to move around the court. When it comes to movement changes, however, horizontal take-offs are an apparent area that is lacking.

Look closely at the return and racket placement – the choice of contact height is higher than most optimal, forcing the player to rely on her upper body, reducing the possible impact her lower body has already put the work in to exert. Closeness of return is however a strength – rather, height of shot is compromised when the player chooses not to move forward upon impact opposed to moving to the ball. The integration of a forwards movement, as with angular, would use the player's body with the shot opposed to remaining stationary and having excess impact applied onto the body, ultimately reducing overall shot impact.

The propulsion of the lower body into the serve isn't quite there. Indeed, the upper body has solid rotation, yet this is problematic when the upper and lower aren't in unison. The upper is quite stiff and has room for further torque – currently the lack thereof and limited upwards push see this player on the backfoot come each respective service game.

A solid all-round player, it's key to be mindful here that although of elite status and accomplishment, it's the choice of height and limited torque that holds her back from climbing deeper into the rankings.

(Ranking, Pre: 46) Matt Ebden

The return of serve finds the player on the backfoot to begin. His ability to cut off the server's angle by moving forward, on an angle, is near obsolete. This becomes problematic over time when the server clues onto his opponent's weakness. To the contrary, the player's serve is solid and gets the job done –

upper body rotation is limited as is the climb up to the ball. In other words, the player near 'dunks' his serve opposed to utilising his whole frame.

The groundstroke mechanics leave a bit to be desired – the lower body is lacking its 'on' switch causing the player to be limited in their control of height of contact. This is detrimental, handing the opponent control of the player's positioning. Yet backswing and elbow height are relatively in toe, creating functional propulsion for the player to use with and against his next point of contact.

Backhand extension is hindered by the player's weakness of moving directionally – this demands use of the slice more often, yet isn't used as frequently as perhaps warranted. The two-handed backhand poses immense challenge when a player is on the run or cannot move to cut off the angle. This is a prime example of player mechanics going out the window and a key opening for any opponent.

The player's use of his lower body is commendable and the most significant factor in his arsenal contributing towards his elite status, currently sitting at a near career high ranking. The overall result, however, is hindered by the player unable to quite take advantage of his current weaker areas to in turn optimise his leading strength during play.

(Ranking, Pre: 46) Ajla Tomljanovic

Upon return of serve the player exhibits flexion in her lower body, yet rises upon contact with minimal flexion remaining. This lift at contact reduces the transfer of kinetic energy and momentum generated behind the ball. Whilst the height of point of contact is relatively controlled courtesy of the player's balanced state, the player does tend to catch the ball slightly late which increases the pressure placed on her upper body. Not afraid of moving around or away from the ball, the player's anticipation is key, yet its effectiveness is limited as a result of her catching the ball higher than most optimal which further restricts the player from extending through her returns without cutting her reach short.

The player's serve benefits from her solid lower body flexion that allows her to increase her upwards lift towards contact. With an arch generated in her back, her lower body counters the pressure, allowing for greater torque to be achieved through rotation, however is limited in its consistent application. Whilst the player is able to generate momentum into the court, this is slight with the player just achieving a post-contact court position. Yet, the player has a near optimal ball toss that allows her lower body positioning to be optimised to reach up and outwards through the ball with her upper body rotation utilising her elastic energy, however the consistency in the toss wavers whereas a more consistent inside court location would see the player further control the outcome of her serve.

When considering the player's groundstrokes her lower body is activated throughout contact, exhibiting an optimal performance indicator. Further, the player maintains a lowered centre of gravity that allows her to use her kinetic energy to her advantage, whilst her upper body exhibits solid rotation. Yet, the player's disadvantage is in her consistency of height of contact, whereas her strength is clear when this position is achieved. When coupled with her movement, to or away from the ball in attempt to control the height of contact, the player uses her vertical movement to her advantage. However, the player's anticipation of the ball at times is slightly late which limits the time for the player to prepare, contributing to her catching the ball either too high, or too close, which restricts the player from gaining an offensive stance.

The player exhibits strong vertical movement in her positioning, not afraid to adjust to or away from the ball. With baseline play favoured, her horizontal movement is consistent, whilst her use of angular movement is also integrated into her game – an optimal performance indicator. Yet, the player's consistency of use in her angular movement and stepping up to the ball is yet to be consistent, although a strength when applied. Whilst the player's momentum through her shots utilises her kinetic energy, whilst her upper body rotation her elastic energy, these transitions remain limited as a result of the player's inconsistency in both height of point of contact and their overall effectiveness, in need of a more consistent use of angular momentum. When the player applies these current inconsistencies, she begins to exhibit qualities of the elite *elites*. Therefore, progressing these current performance indicators into more consistent applications will see this player climb the rankings.

(Ranking, Pre: 48) Dusan Lajovic

When returning serve the player exhibits slight lower body flexion with the tendency to lift at contact leaving the player with limited kinetic energy to transition through his return. Similarly, the player has limited hip flexion at contact that results in a limited lean into the return, compromising his state of balance. Off both the player's forehand and backhand he has a tendency to fall backwards post-contact more often than his forwards transition. Whilst the player rotates through his upper body, he catches the return slightly late at times which keeps him in the defensive position. Likewise, as a result of the player's positioning his height of point of contact is slightly higher than most optimal. The player is limited in his use of momentum through his returns remaining near idle at impact which also results in the player's restriction in extending through his return.

The player's serve utilises momentum to his advantage, at times falling into the court post impact. Whilst the player does use his lower body to drive upwards to the ball through his serve, he tends to keep his shoulders closed post contact that restricts the player from extending through his serve openly. Further, a result of the player's arch, whilst compounded in part by his activated lower body, a result of his limited flexion contributes to the player's less than optimal upwards drive. Similarly, a result of the player's restricted use of his kinetic energy limits the player's vertical lift and at times compromises his landing position post contact.

When contacting his groundstrokes the player does exhibit lower body flexion, however he remains limited in its use or his extension through the return. Further, the player is limited in his capacity to extend through his groundstrokes as a result of both his taller stance and near stationary post-contact positioning. Whilst the player manages to maintain a balanced state through his groundstrokes, more often than not he is consistently placed in a defensive position unable to make the transition to offensive play with his favoured horizontal positioning. To progress, the player is in need of lowering his centre of gravity and altering his technique to allow him to extend through his groundstrokes on both sides, as with heightened awareness of his proximity to the ball, lowering his height of point of contact.

The player's movement favours horizontal positioning with limited vertical or angular movement present. Compounded by the players baseline preference and defensive positioning, vertical movement is scarce. The integration of angular movement would see the player progress in his movement, whilst also his ability to transition to an offensive stance. Further, heightened anticipation of the ball would see the player's timeliness progress and contribute to his ability to achieve an optimal height of point of contact consistently.

(Ranking, Pre: 49) Taylor Fritz

When serving the player exhibits optimal performance parameters that are evident in his end result when his ball toss is in alignment with his extension. Not only does the player draw his lower body towards the ground, but he also rotates through his hips as he pulls back, generating torque through his mid-section, as with his pronounced elbow drop, before releasing through impact. With a service motion of elite calibre, his greatest limitation is in his ball toss to ensure his mechanics can flow through. Near flawless in his mechanics, the player's serve simply is in need of a more consistent ball toss from time, as with his lower body drive that tends to sway at his knees from time as they draw closer, opposed to a smooth streamlined parallel drop.

Upon return the player exhibits slight flexion in his lower body, yet his knees begin to slightly draw together opposed to holding his lower body in alignment to build and apply his kinetic energy through the return. The player often steps into his return prior to impact to use his momentum to compound the speed of the ball. Yet, although the player often flexes at his hips as he rotates his upper body, he is limited in his ability to move with the ball, let alone extend. Further progress on the player's lower body alignment, as with his ability to extend through impact – tied to his proximity to the ball, will allow him to develop a more effective service return.

The player often manages to maintain a lowered centre of gravity through his groundstrokes whilst maintaining an optimal height of point of contact. Often transitioning through the ball, the player is limited in his ability to extend post-impact, often wrapping around to recover, in contrast to finishing through with momentum on his side. The player rotates his upper body whilst keeping his racket low and elbow tucked to his side – an optimal performance indicator.

Whilst the player does in fact exhibit optimal performance parameters through both his forehand and backhand, he is slightly limited by his willingness to extend through the ball – at times caught muscling the ball with his upper body as a result of lifting too soon. Although minor, the player's proximity to the ball, with slight adjustments, will enable the player to more readily extend through contact whilst limiting the pressure placed on his upper body.

As a baseline player, horizontal movement is relatively streamlined for the player as he endeavours to maintain his positioning. From time the player moves vertically as he adjusts to the oncoming ball, slightly up to the baseline or in front, yet the player tends to maintain his position on the baseline in contrast to moving further forward. With angular movement scarce from the player's game, a purposeful integration would begin to bring his game closer together whilst allowing him to take a more offensive position through play.

(Ranking, Pre: 50) Viktoria Kuzmova

When serving the player activates her lower body that is accompanied by the player dropping her racket near-vertically by her side before beginning her extension up towards the ball. With good upper body rotation that allows the player to make the most of her elastic energy, the player often places her ball toss inside the court to help release through impact and take advantage of her momentum. With hip flexion evident post impact, the player has near optimal technical parameters through her serve. To progress, the slight lag in her racket drop stalls the transfer upwards of her kinetic energy – a slight improvement on this fluidity is key to take advantage of the player's release through impact.

As the player returns serve she keeps a low base of support as with an optimal height of point of contact as she leans with the ball through impact. Although limited in her transition through the return with a near stationary position post impact, courtesy of her lean behind the ball, at times moves to take the offensive position. With good anticipation of the serve, in particular when it falls short, the player rotates through her upper body and releases through the ball whilst keeping her optimal performance parameters. The player has a tendency at times to wait for the ball and catches is slightly higher – greater consistency with her height of impact will see the player dictate play more often.

The player often exhibits optimal performance parameters through both her forehand and backhand whilst maintaining the presence of dynamic balance through impact. With groundstrokes that near the calibre of the elite *elites*, the player's current weakness is in her consistency of height of point of contact. Further, for the player to consistently extend through her groundstrokes, as with the ability to use her momentum through impact, opposed to her often near stationary position post impact, will see her deliver against the elite *elites* more often.

Often positioned on the baseline, the player moves horizontally with ease, whilst using vertical movement as she steps up at times, or backtracks, to endeavour to maintain her optimal height of point of contact. From time the player will move angularly, however the player's integration of angular movement in scarce. For the player to progress and join the elite *elites*, angular positioning is key as with her ability to extend through impact.

(Ranking, Pre: 51) Sam Querrey

Upon return the player slightly activates his lower body with a forward lean at the hips before further lowering his centre of gravity at impact with a controlled height of point of contact. Exhibiting optimal performance parameters through his return, the player is limited in his forwards transition, however does step up to the ball often prior to contact. At times the player is caught with a higher than optimal height of contact, varying in his consistency. With an abbreviated swing – an optimal performance indicator, the player rotates his upper body just enough to complement the dynamics of his body to compound the force behind the oncoming ball – another optimal performance indicator. Whilst the player does exhibit a return of the elite *elites*, return of height of contact and his consistency, as with his forwards transition, are key to progress in the ranks.

The player often exhibits optimal performance parameters through his groundstrokes – a controlled height of point of contact in sync with a lowered centre of gravity through the ball. With a tendency to stop short post contact, the player does manage to rotate through his hips – using torque to his advantage as with the presence of dynamic balance. Whilst he does rotate his upper body to transition through the ball, he tends to be more closed off on his backhand, limited in its extension, yet his forehand does manage to extend more often through impact. Although the player is limited in his forward transition, the consistent presence of dynamic balance sees the player counter the oncoming ball with a foot-back approach post impact, before letting the left foot back down – ensuring his centre of gravity remains balanced. To have greater impact, the player's use of angular movement to cut off the angle is limited, tied to his extension and momentum through his shots, as with his consistency of height of contact – integrating these performance indicators would allow this player to progress even further.

When serving the player uses his longer levers to his advantage, exhibiting quite often, optimal performance parameters through his serve. With flexion apparent in his lower body, coupled with a fluid upwards drive and inside court ball toss, the player near-flows into the court post impact. This smooth and fluid service motion is further excelled by the player's levers which enable him greater height on the ball which affords a higher angle of impact to generate force through the ball. With the arch in his back supported by the flexion in his lower body, although slight, the player's natural flow and use of his levers are of the elite *elites*. With his upper body rotating as with his hips through impact and ultimate transition into the court, this player showcases optimal performance parameters of the serve. There is room for an increased range of flexion in the player's lower body which would result in greater kinetic energy made available to transition upwards and through the ball.

With a preference for baseline play, the player opts to move horizontally most frequently with limited use of angular positioning. Whilst he isn't shy of moving vertically when moving in behind a shorter ball, or serve from time, the baseline is where he tends to remain. A result of the player's scarce use of angular movement sees the player often caught on the defensive and allows his opponents to keep the offensive position. However, by the player integrating angular movement specifically would not only allow him to cut the angle off the oncoming ball, but more readily change his defensive positioning to a more offensive stance to begin to more often dictate play.

(Ranking, Pre: 51) Victoria Azarenka

When serving the player achieves slight lower body flexion before extending upwards towards contact. The slight limitation in the player's lower body flexion hinders the upwards drive to and into the court and the resultant momentum behind the ball. Whilst good upper body rotation is present, the player does linger between her racket take back and elbow drop ever so slightly, impacting the player's service fluidity and release behind the ball. As a former elite *elites* with clear optimal performance parameters, the player's lower body is merely limited in driving the player upwards more prominently as her post-impact position is somewhat strained.

The player's groundstrokes primarily exhibit optimal performance parameters as she maintains deep flexion in her lower body coupled with her controlled and optimal height of point of contact. Yet, the player does lift from time, more often on her forehand than backhand, that limits the impact of her ball as with the transition of her kinetic energy. Further, whilst the player moves with her groundstrokes, she also has a tendency to keep her position. Therefore, whilst optimal performance parameters do shine through, the player does have phases where she stands a little more taller, where her lower body is slightly lifted, and where she is limited in her extension through impact. As a former elite *elites*, the player's current weakness is in her consistency which is also tied to her limited use of angular positioning – a key performance indicator she's in need of further integrating to heighten her ranking.

Upon return the player gets down low with a prominent forward lean into the court as she activates her lower body. With the tendency to lift from deep flexion to slight flexion through contact, at times, the player primarily manages to maintain an optimal height of point of contact. Yet, when the player maintains the depth of her lower body she places herself immediately on the offensive more often than not. The player exhibits quite frequently optimal performance parameters through her return, from her forwards lean, extension and transition through the ball, as with her optimal performance indicators in her lower body and height of contact. As a former elite *elites*, this is no surprise. The player's current weakness is evident when she lifts through contact, forced into a defensive position.

A versatile mover, the player moves both horizontally and vertically with ease and often will move vertically as she re-positions herself to maintain the optimal height of point of contact. With evident optimal performance parameters throughout her game, the player's current weakness is merely in her consistency of application. A primary factor the player is yet to consistently integrate is her use of angular positioning, often maintaining her baseline position, or behind, as she hesitates to: re-position herself with her hips facing inside the court to allow her to not only rotate through her midsection, but to cut off the angle of her opponents whilst further stepping up to the ball to keep a consistent height of point of contact throughout play.

(Ranking, Pre: 52) Sofia Kenin

Upon return the player slightly flexes her lower body as she has a tendency to make contact with the ball somewhat higher than most optimal. A consequence of the player's higher point of contact results in the player having limited lower body activation through contact as her kinetic energy is dispersed upwards more often than through the ball. From time the player will manage to make contact lower – inside the optimal range of impact, yet is limited in her consistency. The player often impacts the return closer than most optimal that restricts her from extending through the ball, as with a taller than optimal upper body stance. Flexion at the hips as with progress on her proximity to the ball, and a more consistent height of point of contact, will see her return most effective against the elite.

As the player serves she slightly drops her centre of gravity before extending towards the ball – limited in her upwards drive and extension through the ball as a consequence of the player's often near-inactive lower body as she transitions through impact. Yet, the player does manage momentum behind her ball, although slight, as a result of her inside court ball toss. To further advance her serve the player is encouraged to build on the activation of her lower body to not only generate an effective upwards drive for the player to extend through impact, but also to optimise her current technical parameters and to progress in the ranks.

The player maintains a lowered centre of gravity through both her forehand and backhand as with an optimal height of point of contact more often than not. Whilst the player's proximity to the ball is slightly too close, inhibiting her from extending openly through impact, the player's slight flexion at her hips sees her pivot forward, from time – an optimal performance indicator. At times that player does make contact with her groundstrokes outside the optimal range for contact that results in the player lifting through and post impact. The player's weakness is her ability to transition through her groundstrokes and into the court, most frequently making impact followed by opening her hips without the use of extension, or momentum. To take her game to the next level, the player's proximity to the ball is key to better gauge her height of contact, whilst also encouraging the player to naturally transition through her groundstrokes, using momentum to her advantage.

Most often the player moves horizontally along the baseline, limited in her vertical movement. When presented with a shorter ball the player will move towards it, however scarcely will the player adjust her baseline positioning, forwards or backwards, although slight, to better control her height of contact. Whilst an avid mover, her positional movement is yet to be integrated into her game with angular positioning rarely applied. A deeper and more purposeful use of angular movement will benefit this player greater with her already clean technical parameters. Further, by applying angular positioning through play the player will be in a more effective position to extend through impact and consider a more consistent height of contact coupled with a deeper level of flexion through her shot.

(Ranking, Pre: 53) Monica Puig

When returning serve the player exhibits lower body flexion pre and post impact – an optimal performance indicator. Likewise, the player adjusts her positioning to meet the ball at the optimal height of point of contact, although at times does let it rise slightly outside this range which places added pressure on her upper body as she endeavours to compensate for the added force unable to be compounded by her lower body. The player is limited in her transition through the ball when returning, as with the lack of momentum. Whilst the player does rotate through her returns with an abbreviated swing, she does have a tendency to wait for the ball at times opposed to moving angularly to cut off the angle of return. Although slight, progress on the player's current technical parameters will bring her back to the elite niche.

Through her serve the player activates her lower body and further generates torque prior to extending up and through the ball – an optimal performance parameter. Further, the player rotates her upper body and extends in sync with the drive from her lower body. Yet, the player’s serve is slightly limited at times from her ball positioning whilst technically she continues to exhibit optimal performance parameters. Although the player doesn’t have the fastest serve, technically she has the capacity to generate further force behind her serve with an increased upwards drive – slightly, as a result of the player already achieving good activation. A move in the player’s ball toss, however, will promote a more inwards lean and naturally allow her body to fall behind the ball.

The player's backhand and forehand apply optimal performance parameters. Not only does the player move slightly forwards, or backwards, to adjust her height of point of contact, but she maintains an activated lower body throughout, as with the presence of dynamic balance. With indicators of the elite *elites*, the player generates torque post contact that sees the player exhibit slight momentum behind her balls. However, the player's weakness is in her consistency – at times opting not to move with the ball as she gets stuck in her positioning and then lifts through contact. Yet, the player does primarily execute her groundstrokes most optimally and is typically inconsistent at times as a result of her positioning.

The player applies angular movement more often through her shots as she is capable of generating torque and releasing through impact. Further, whilst horizontal movement is favoured, the player isn't shy of stepping up at times to cut off the balls angle – although this isn't as frequently applied as necessary. When presented with a short ball, the player is comfortable in her vertical movement, yet does typically stay back on the baseline. A more pronounced integration of angular movement through play would see the player extend on her offensive stance, whilst also work in her favour to finish her extension through her groundstrokes opposed to her often early finish, as with her momentum behind the ball.

(Ranking, Pre: 54) Belinda Bencic

The service return of this player is slightly amiss – from the lower body sitting too high, and contact height higher – a direct consequence that sees the player swinging her arms on return opposed to utilising her lower body. Similarly, although balanced her centre of gravity sits too high which forces her upper body to take the burden of the shot – a common limitation for player's who haven't developed their lower body to perform as optimally required.

When it comes to the service motion there are similar unconditioned elements that are yet to be used to the player's advantage. Her lower body is used, however the range of motion is limited. Likewise, upper body rotation is scarce. The player gets away with her service motion by having a smooth technique – yet limited by her lack thereof of optimal technical parameters that hold back such advantageous proponents of play such as power, energy and momentum.

A similar case is noticeable in her groundstrokes – sitting slightly too high and consequently making contact equally too high placing an undesirable load on her upper body. Further, as a result of this point of contact, the player's centre of gravity sits higher than optimal limiting her state of controlled balance whilst neglecting the input of her lower body. Ultimately, it's an upper body game for this player as she navigates the more optimal use of her lower body.

Keeping her in the game is her ability to move through her shots and into the court when desired. This level of consistent moveability keeps her in the game. Although it isn't as functional as desired nor does the player utilise the strengths of her body, an offset of this is the heightened amount of overall strain for her body to take the brunt. To counter this, it is key that the player begins to utilise her lower body and begin countering the forces placed on her joints by increasing the application of functional movement into her game.

(Ranking, Pre: 54) Matteo Berrettini

When serving the player endeavours to use his upper body to achieve added momentum through his serve, bringing his arm swing to a back-forth motion as he activates his lower body before extending upwards to the ball. This slight off-rhythm of the player's service motion doesn't necessarily hinder his service outcome, yet it does look to affect the overall fluidity of the player's service motion. Although the player does rotate through his upper body and extend upwards to the ball, his outer hip opens slightly too soon at times. Further, the player's ability to achieve momentum behind his serve is hindered at times by the inconsistency of his ball toss. Whilst the player has a technically sound serve, slight variances in player service rhythm does have a tendency to displace the overall service mechanics, leaving it more open to error. To progress and achieve greater consistency and effectiveness through his serve, the player is encouraged to address his rhythm and look towards a more pronounced lower and upper body synchronisation without the potential rhythmic distractions, as with an increase in flexion in his lower body to offset his at times pronounced arch in his back.

As the player returns serve he activates his lower body whilst his upper body primarily stays vertical as he is limited in flexion at the hips. The player varies his height of point of contact with an inconsistency noted here, whilst his lower body has a tendency to lift as contact to limit the level of flexion present at impact. These limiting factors contribute to the player unable to transition and extend through his return's consistently. When receiving the ball, the player is limited in his adjustments prior to impact which results in him being caught at times too close to the ball. To progress, the player is in need of adjusting his proximity to the ball to not only allow him to contact out in front, but also to extend through the return opposed to getting caught in a closed position, as with the need to flex at the hips to lean into his returns.

The player manages to lower his centre of gravity when contacting his groundstrokes as he also rotates through his hips with good upper body rotation. However, at times the player restricts his upper body rotation on his forehand, keeping an abbreviated swing pattern, whereas on his backhand the player is more fluid. With a height of point of contact more optimal than not, the player is let down at times from his upper body positioning – standing tall opposed to leaning into his shots. Whilst the player is consistently balanced throughout play, he does at times fall off his shots or keep a stationary position – rarely extending through his groundstrokes. Further, the player's anticipation of his groundstrokes at times, as with his fall-back post-contact, opposed to moving with his groundstrokes, hinder his ability to progress and maintain an offensive stance.

Whilst the player favours baseline play and moves horizontally quite often, he does apply vertical movement through his game. Perhaps the player's biggest limitation is his ability to further integrate the use of angular movement through play. Whilst he has his noted weaker areas, the player can readily address these by a more purposeful use of angular positioning through his game. Not only will this new positioning encourage the player to fall into his shots, but help him extend through his balls whilst also take time away from his opponents – a starting point to move up in the rankings.

(Ranking, Pre: 58) Dayana Yastremska

When serving the player lowers her centre of gravity that allows the player at impact to transition into the court, using her momentum. As the player extends towards the ball, her technique is fluid as with the location of her ball toss. However, at times the player's ball toss is slightly closer to her body, restricting the player's ability to transition smoothly into the court. Whilst the player exhibits optimal performance indicators through her serve, the player's weakness is merely in her consistency and at times, depth of her lower body coupled with the proximity of her ball toss.

Upon return the player manages to stay low to the ground as she extends through the ball – courtesy of her controlled height of point of contact, however this is inconsistent. More often the player has a tendency to lift prior to impact to meet the ball above the optimal height of point of contact whilst her kinetic energy is transferred upwards, opposed to through the ball – a consequence of her upright stance. Whilst the player rotates through her upper body, she is limited in her ability to extend and transfer through the ball, often caught in a near stationary position post contact simply waiting for the next ball – contrast to committing to her return and then using angular positioning to her advantage.

The player exhibits optimal performance indicators through both her forehand and backhand, however is inconsistent in their use. Often the player will lift through contact as her lower body extends upwards an she is reliant on her upper body to deflect or take the brunt of the oncoming ball – contrast to applying her stored kinetic energy against and through the ball. However, when the player does manage to exhibit her at times optimal height of point of contact, tied with her evident flexion in her lower body, not only does she take the offensive position, but more often transitions through the ball. Whilst the player is limited in her ability to extend through the ball openly, she does from time transition through both her forehand and backhand. With a tendency to pull off the ball and recover to the centre, the player limits her use of momentum behind the ball. To progress, although the player does exhibit elite aspects of her game, she is held back by her consistency. More often the player's current weaknesses can be met with a higher level of consistency through play, made simple with the use of angular positioning.

Often moving with ease around the court, the player's preference for baseline play see's horizontal movement more often than vertical movement. Whilst the player does use her vertical movement to move up on shorter balls, she often uses her horizontal movement as she is limited in her ability to step up through play. Further, the player is yet to integrate angular movement into her game, holding her back from not only extending through her shots, but taking time away from her opponents as well as using it to her advantage as she sets for the oncoming ball and transitions through. Whilst the player does exhibit optimal performance indicators from time, further development of her game is clear, as with her consistency through play.

(Ranking, Pre: 59) Timea Babos

When serving the player is primarily technically sound, yet does have the tendency to open her hips too soon to be in a near front-on position from time as she extends towards the ball. Further, the player has the tendency to limit her depth of flexion in her lower body that results in the player leaning back as she drops her elbow, in contrast to her lower body countering the pressure of her back to allow her then to release up and through the ball. However, when the player does increase the depth in her lower body, her service motion tends to be smoother as she opens into the court, although scarce. To progress, the player is encouraged to lower her depth of flexion to offset her arch, whilst holding back before she releases her hips as she impacts through the ball.

As the player returns serve she exhibits slight flexion in her lower body with slight flexion at her hips, although does tend to stand taller in her upper body through contact in contrast to leaning through and with the ball. Using an abbreviated swing often through her return, the player rotates her upper body to hit through the ball, yet the player is limited in her transition through impact as with her extension through the return. However, the player does lower her base of support through impact at times that allows her to lean into her return. A more consistently applied technique through her return, primarily a more controlled height of contact, will begin to further develop the player's return.

The player more often exhibits an activated lower body through her groundstrokes – an optimal performance indicator. Yet, when the player makes contact with the ball she has a tendency to make contact slightly above the optimal height of point of contact that keeps her from transitioning through her shots. Limited in her ability to extend through her groundstrokes off both sides, the player does use her momentum through impact from time as she releases through her hips. Height of point of contact aside, the player does exhibit optimal performance parameters quite often through her groundstrokes, whilst she is held back by her height of impact and occasionally her proximity to the ball that restricts her ability to extend through impact.

With a preference to maintain her baseline positioning, the player comfortably moves horizontally through play, whilst from time will step up, or away from the ball as she moves vertically. The player does integrate angular movement through play, however this is scarce and is in need of a deeper integration into the players game. To progress, the player is in need of stepping up slightly more to the ball to control its height of point of contact, or backwards, to ensure she is able to meet the ball where it's most optimal to allow her to then extend through impact and look to place herself in a more offensive position during play.

(Ranking, Pre: 62) Ryan Harrison

When returning serve the player activates his lower body and flexes slightly at the hips to maintain a forward lean into the court. This alone exhibits a controlled centre of gravity and the player is rewarded with a consistent state of dynamic balance. The player's ability to rotate his upper body whilst staying low to the ground keeps him on an even toe – not forcing him into a state of defence. Further, by the consistent use of his lower body and not afraid to reposition pre-contact, he has the ability to control his height of contact when he chooses – a slight weakness is indicated here with this inconsistency.

The player's serve achieves lower body flexion and solid upper body rotation, however the positioning of his ball toss causes a greater arch than supported by his level of flexion. The player's current sequence slightly halts his momentum and upwards drive to push out into the court. This is exhibited through the player's limited forwards momentum with the player relying on the rotation of his upper body to hit through the ball. To the contrary, a result of the player's current sequence, although with its limits, is his ability to achieve forwards momentum into the court as a result of his lower and upper body remaining in sync. Perhaps the greatest limitation here stems from a deeper lower body drive, inconsistency with his ball toss and its technical merit, that results in the players current outcomes – attention to these areas, although slight, would place this player's ranking in higher contention.

The player's groundstrokes maintain a solid base of support with solid upper body rotation. Perhaps the players greatest limitation is his proximity to the ball at contact – slightly too close to allow for that smooth transition and reach through the ball. Interestingly, when the player's movement is in sync he does in fact allow himself room to transition through the ball – consistency of placement and readiness here is key. The player's height of contact through his groundstroke is higher than most optimal, more often than not – lowering this back into a controlled state coupled with his ball proximity would allow this player to further excel.

Horizontal, vertical and angular movement are all within the player's repertoire, yet his level of anticipation and his ability to subsequently react are limitations. At this level of play this is merely slight, as overall the player has a near elite game. Highly accomplished in his own right, it's the little things that set the elite *elites* from the elite, and in this player's case he has a solid skillet and is simply missing that extra level of depth, speed and readability.

(Ranking, Pre: 63) Denis Kudla

A mechanically sound serve, the player exhibits a good arch that is supported by the flexion in his lower body. When it comes to his upper body, although the player achieves a level of rotation, he is somewhat limited in its transition through and into the ball. This is mostly limited by the player's lower body upwards lift – even though flexion is present, the release of his kinetic energy and its transition upwards is restricted that in turn sees his upper body more cautious to open and release its elastic energy. A key performance indicator is the player's ball toss that although is out in front, his momentum stalls. To see the player take advantage of his kinetic energy and then transition into the court, the ball toss can be slightly moved further in front to encourage that slight fall into the court.

When returning, the player manages to activate his lower body, yet prior to contact does lift, compromising his level of dynamic balance and base of support. This change in the location of his centre of gravity sees the player more often than not pushed back and into a defensive position opposed to moving with his return. The limitation in momentum is also a result of the player's near stationary position at contact, being dictated by the ball opposed to simply maintaining his balance and flexing at the hips to compound this near reflex action. Likewise, the player's height of contact is often compromised as a result of maintaining his position and the limitation in his movement.

The player has a relatively consistent base of support through his groundstrokes, however due to his favoured horizontal movement, is limited in his transition to offensive play. Often caught behind the baseline when a shorter ball appears, vertical movement is restricted due to his already behind positioning. Although the player is technically sound with good rotation on his backhand, whilst stepping through his forehand more often, the outcomes are more often not detrimental to his opponents as a result of the player's limitation in moving into the court and taking a more offensive stance. The player is further limited by staying back in controlling his height of point of contact, often not on his terms which sees his base of support often compromised whilst limiting his ability to extend through his shots.

A clear factor that if integrated would see this player progress in the ranks is his ability to move forwards and the use of angular movement throughout his game. This coupled with his ability to utilise and transition his momentum through his shots are key performance parameters that are in need of application tied to his level of anticipation to respond more appropriately, opposed to staying back, to the oncoming ball.

(Ranking, Pre: 66) Stan Wawrinka

The player near sways his upper body, forward and then back, as he prepares to make impact with his ball toss. Upon swaying back he flexes his knees in sync with his back – likened with a backwards sway motion before rotating his hips and extending upwards, the player comes forward to extend through impact. The player uses this sway as a way of generating momentum behind his serve, whilst also the release of his elastic energy. Whilst the depth in his lower body is at times slight, he does also prolong this depth to generate a greater build in his kinetic energy. Technically, the player's serve varies from a more orthodox service motion, courtesy of the player's unique sway through impact, allowing his serve to challenge the elite niche. Greater consistency in the player's lower body depth as with his flexion at the hips will progress this player back to where he once was – inside the elite niche.

When returning serve the player leans into the court as he flexes at his hips, exhibiting pronounced flexion in his lower body – an optimal performance indicator. Using an abbreviated swing through contact, the player often moves slightly forward with the ball as he transitions through. Near consistently achieving an optimal height of point of contact through his returns, the player's use of his kinetic energy against the ball sees his returns more effective than not. The player's weakness is his upper body stance at times, lifting early at the hips that results in the player's upper body taking the pressure opposed to moving through with his body. Yet, more often than not the player exhibits qualities of the elite *elites* with his extension through impact a primary indicator of the elite niche.

Through the player's groundstrokes he manages to keep a lower centre of gravity with pronounced flexion through his single handed backhand, as with his extension. The player equally stays low on his forehand and often extends through impact as be transitions off both sides as he uses momentum behind the ball most often than not. When the player is caught off the baseline he is however more likely to raise his height of contact at times as with when he transitions through the ball, lifting his lower body in contrast to maintaining his use of kinetic energy. Yet, the player effectively uses his momentum, as regardless of position, manages to extend through the ball, frequently exhibiting qualities of the elite *elites*. With optimal performance parameters apparent though the player's groundstrokes, showcasing the calibre of play that once had him inside the elite *elites*, the player's primary weakness is in his use of angular movement more often as he anticipates the oncoming ball, at times often getting caught moving horizontally in contrast to stepping up and dictating play.

As a former elite *elites*, the player's movement is key to align with his optimal performance parameters throughout his game. With horizontal movement seamlessly integrated, his horizontal movement doesn't have the 'spring' it once did, limiting his effectiveness in moving further inside the baseline at times. Yet, the player adjusts his stance vertically behind the baseline at times near effortlessly, courtesy of his consistent presence of dynamic balance. A primary performance indicator scarce in the player's game from where it once was, is his use of angular positioning. Although integrated from time, its consistent application is not as frequented as in need, limiting the player's shot effectiveness through play, in particular when on the run, often placed on the defensive. A consequence of the player not positioning himself angularly, or moving up on the angle to cut off the angle of the ball, gives his opponent the advantage in these instances and is a primary limitation halting the player from joining the elite ranks from where he once was.

(Ranking, Pre: 67) Aljaz Bedene

Upon return the player flexes his lower body slightly with limited flexion at his hips. As a result of his taller position, the player makes contact with the ball higher than most optimal with limited transfer of kinetic energy behind the ball. The player rotates through his upper body with limited transition through the ball as with his momentum. Although technically sound, the player's limitation in stepping up and extending through the ball is further compounded by his higher than optimal point of contact as with his limited flexion through his lower body, more often through impact, dispersing his energy prior to making contact.

When serving the player manages to lower his centre of gravity, yet his upwards drive is limited as he reaches for the ball with a limited upwards lift as he goes to extend out and through the ball. As the player transitions through his serve he maintains a taller than ideal upper body which holds back his momentum into the court. Whilst the player does achieve good upper body rotation and a drop with his elbow, his ability to propel himself upwards and into the ball is restricted by the work in his lower body. Increasing the load in the player's lower body to contribute to a more significant upwards lift would allow the player to work towards a more effective serve.

Through the player's groundstrokes he manages to activate his lower body through contact and often achieves an optimal height of point of contact. Whilst the player appears to exhibit optimal performance parameters through his groundstrokes, he is limited in his consistency and primarily, his transition through the ball – often caught with a tall upper body. As a result of the player's near-upright upper body, at times the player inadvertently places the pressure on his upper body in contrast to using his kinetic energy to his advantage and transitioning through his shots – forwards. Further, the player's execution of his optimal performance parameters do become inconsistent as he tends to rise through impact, again, displacing his ability to apply increased pace on the ball and taking the offensive stance – most often caught on the defensive. Further work on the player's consistent stance through impact, as with his ability to extend through his groundstrokes will see him progress with the integration of angular positioning.

The player tends to use horizontal movement through play as he opts to stay on the baseline most often. Although the player will move vertically when given the shorter ball, he is limited in its use as he near pulls back on his groundstrokes, hesitant to transition forward. A result of the player's limited use of angular movement is the player covering a wider distance on the baseline – unable to cut the angle off his opponents ball by stepping up into the court and angling his hips to face inside the court. To progress, the player is in need of a more prominent use of angular movement in his game to look towards taking the offensive position, whilst also working on his consistency as he extends through his groundstrokes.

(Ranking, Pre: 68) Andrey Rublev

Upon return the player activates his lower body and isn't afraid to step up to the ball and apply angular momentum. With flexion at the hips, the player manages to slightly lean into his return and transition through, more often on his forehand opposed to his backhand. The player controls his height of point of contact for the most part, however lifts at contact to be left with limited lower body activation and thus kinetic energy to transfer through his return. Whilst upper body rotation is present, as with the player's extension through the ball, momentum behind his return is limited post impact, often keeping the player on the defensive. To progress, the player is primarily in need of a more activated lower body and willingness to transition through his returns to use his kinetic energy to his advantage, furthered with a more purposeful use of momentum.

The player exhibits the use of momentum through his serve furthered by solid flexion at the hips, although wavers in his consistency. With an activated lower body, the player tends to sway sideways upon impact opposed to a smooth transition upwards and outwards through the ball. It becomes apparent that the player is in need of moving his ball toss slightly inside the court, with a more vertical lift through his serve. Further, with rotation apparent in the player's upper body with a sound technique evident, the player's most significant limitation is tied to his lower body lift up to the ball, coupled with the location of his ball toss. Progress on both limiting factors will see the player begin to near optimal performance parameters.

When contacting his groundstrokes the player predominantly manages to exhibit flexion in his lower body with a moderately controlled height of point of contact. Not afraid to run around his backhand, the player uses his agility to his advantage with timeliness on his side. Whilst good upper body rotation is present on both his forehand and backhand sides, the player has a tendency to pull back on his backhand from time. Further, although the player activates his lower body, he has a tendency to lift at contact from time with a higher than optimal centre of gravity. Yet, when controlled the player is at his best. Similarly, when the player raises through his groundstroke's errors begin to spark, however with dynamic balance on his side more often, the player has the ability to progress towards optimal performance parameters with these adjustments in his application and consistently through impact.

A favoured horizontal mover, the player is often caught behind the baseline and scarcely moves vertically. Yet, the player at times does integrate angular movement into his groundstrokes to cut off the angle of the oncoming ball – an optimal performance indicator. Whilst the player is limited in his consistency in keeping his centre of gravity low to the ground, he manages to counter this at times with the presence of dynamic balance – yet, this continues to halt the player in taking an offensive stance. To progress, the player is in need of maintaining a lowered base of support through his shots to benefit from his mostly controlled height of point of contact, coupled with a more pronounced use of his angular movement to see him move through the rankings.

(Ranking, Pre: 70) Yafan Wang

Through her return the player activates her lower body, although often gets caught with a higher point of contact. The player does manage to transition through her returns, yet is limited in her ability to openly extend and use her momentum. At times the player steps up as she goes to return, achieving an optimal height of point of contact. Whilst the player exhibits optimal performance indicators from time as she returns, her weakness is in her consistency of application and their use in conjunction with one another. Further development of these performance indicators, as with the player's ability to better gauge her proximity to the ball, placing her in a more optimal position to extend through impact, is key.

As the player goes to serve she achieves a pronounced level of flexion in her lower body that she uses effectively to propel her upwards towards the ball. With a ball toss inside the court, the player generates momentum post impact that pushes her inside the court. The player exhibits a clean extension through impact courtesy of her upper body rotation in sync with her upwards drive, however from time brings her back leg forward early post-impact, indicative of the player restricted in her transition through the ball and the rotation through her hips, placing a heavier reliance on her upper body whilst halting the player's momentum behind the ball.

The player maintains a relatively consistent lowered centre of gravity through both her forehand and backhand groundstrokes whilst frequently exhibits an optimal height of point of contact. Whilst the player does identify with optimal performance indicators through her groundstrokes, her weakness is in her consistency of application as with her ability to extend through both shots. Yet, the player does achieve hip flexion through contact as she leans into the ball, helping her transition through the shot – although she holds back on her post-impact momentum. The player does generate torque through her groundstrokes as she rotates through her hips, however is yet to integrate angular positioning to be apparent through her game.

The player tends to maintain her baseline positioning with horizontal movement used most often, whilst vertical movement is limited to shorter balls. Whilst the player does make slight adjustments prior to impact, a result of her limited vertical movement sees her catch the ball high – outside the optimal height of point of contact. With angular movement scarcely identified through the player's game, progressing to integrate this form of movement is key to her advancement which will further work with the player to help her better control her height of impact.

(Ranking, Pre: 71) Tomas Berdych

Upon return the player activates his lower body, however has a tendency to limit his lower body flexion at times restricting his ability to transition through the ball at impact. Primarily, the player has a controlled height of point of contact as he also transitions behind and through his returns. Using an abbreviated swing as he rotates his upper body, the player from time will step up to his return whereas his limitation is in his willingness to move angularly to cut the angle off his opponents serve. Whilst the player has a smooth technique as he returns serve, his weakness is in his transition through the ball and at times his closeness on his backhand.

When serving the player activates his lower body which offsets the arch in his back as he reaches to extend towards his ball toss – primarily located inside the court. With momentum achieved post contact, the player achieves optimal upper body rotation as with his elbow drop to release his elastic energy through impact. The player's lowered centre of gravity through his service motion in sync with his upwards drive is an optimal performance indicator as with the player's smooth technique. At times the player occasionally sways as a result of his ball toss and the player's lower body not activating as deeply – this places added pressure on his back as his lower body isn't as prepared to offset his service pressure. However, this is an inconsistent part of the player's game that comes down to consistency in his service rhythm through play.

The player manages to maintain a lowered base of support through his backhand as he more often extends through his shot with an optimal height of point of contact. Through the player's forehand he does have a tendency to catch the ball late, at times, however when he meets it in front to extend through, he exemplifies optimal performance parameters. As a former elite *elites*, the player manages to show signs of his former ranks – yet, his greatest limitation is his timing as with his consistency. Affording himself slightly more time to set through his shots, as with his ability to transition through – often stationary post-impact or performing his groundstroke with a slight pivot, keeps the player on the defensive opposed to stepping up to dictate play.

With a preference to stay on the baseline, the player moves horizontally with ease, yet his willingness to transition through his shots or move vertically is limited unless a shorter ball comes his way. Similarly, the player is scarce in his angular positioning. Whilst he does exhibit signs of the ability to rotate through his hips, often the player neglects to apply an angular stance that sees him restricted in his ability to then take an offensive stance. Further integration of angular positioning is key, as with the player's use of momentum through his groundstrokes in particular to begin to progress to the calibre of play he once mastered.

(Ranking, Pre: 72) Jordan Thompson

The player's service motion is fluid with a smooth flow between his lower and upper body before reaching for impact. Whilst hip flexion is present, accompanied by an arch in his back, his lower body manages a good range of flexion to offset the load placed on the player's back. Yet, the player has an identifiable weakness in the placement of his ball toss – moving it slightly forwards into the court will progress the player with his momentum into the court. Coupled with increasing the speed of rotation between the player's upper body take back and upwards swing will see the player secure a more effective optimal service motion.

When considering the player's groundstrokes it is notable that his height of point of contact is inconsistent, although his lower body does remain activated to varying degrees. A result of this inconsistency is the player having varied heights of contact opposed to moving more upwards and angularly to the oncoming ball to control his height of impact. Interestingly, the player catches his backhand slightly late whilst his forehand hits through the ball more often; both sides are limited in the player's ability to transition with and through the ball, completing the extension. Further, whilst the player has a smooth slice backhand that is also used during play, a result of his limited angular movement typically sees the player attempt to use topspin at times that pulls him off the court – integrating angular movement to his game will allow these areas to be progressed.

Upon returning the player maintains a low and balanced stance, yet has the tendency to fall back at impact – opposed to moving with and through the ball. The player does step into his returns from time, however is limited in his consistency. This is mostly restricted as a result of the player waiting for the serve to come to him which further compromises his height of contact at times – it becomes evident that consistency is a limitation for this player and slight progression with his proximity to the ball will further this player's overall results.

With limited angular movement coupled with a closer proximity to the ball, restricting the player's ability to transition through the ball – are key performance indicators for progression. Yet, as a solid all-round player the limitation is a result of his favoured horizontal movement over his vertical and angular momentum used to his advantage – holding the player back from joining the elite ranks.

(Ranking, Pre: 72) Tatjana Maria

When returning the player slightly lowers her centre of gravity that results in the player having a height of point of contact slightly higher than most optimal. Yet, when the player does manage to stay low to the ground her ability to hit through the ball is more consistent. With a tendency to slice her forehand and backhand as she returns, staying low is key – whilst her height of point of contact is more often controlled here, as with her lower body, yet the player scarcely moves forward through the return to use her momentum behind the ball. To progress, the player is in need of a more consistent base of support and lean into the court, as with her ability to extend through her returns before setting for the next ball to enable her to develop a more controlled and effective service return.

As the player serves she flexes her lower body as she extends upwards to the ball. Yet, a result of the player's limited lower body flexion sees her upwards drive limited. A direct consequence of this limitation is a restriction in the transfer of kinetic and elastic energy through the ball with a relatively slow service speed in contrast to the elite pace of ball speed. Although she does rotate her upper body, the placement of her ball toss limits her extension through the ball which is further limited by the player's higher than most optimal range of movement at her knees. The player attempts to arch her back as she reaches towards the ball, however due to the limited flexion in the player's lower body, this is restricted due to the player's near upright position as she extends to the ball, leaving her with near null energy to transfer behind her ball – a direct consequence of her tall positioning.

The player lowers her base of support through her groundstrokes, slightly, limiting her ability to consistently control her height of point of contact. Whilst the player endeavours to maintain a lowered centre of gravity, a result of her limited anticipation of the oncoming ball sees the player often maintain her positioning, reliant on contacting the ball in her given position. However, with the use of her slice off both sides, the player is capable of lowering herself towards the ground as with maintaining a lowered height of point of contact – yet this is reliant on her positioning which is a current weakness that affects the outcome of her groundstrokes. Further, the player is limited in her extension through the ball. At times the player will hit through, though her forwards momentum is conditional to whether the player intends to approach the net. To progress, the player is in need of a more consistent base of support as with her anticipation to better gauge her proximity to the ball and ultimately her height of point of contact to look towards extending through her balls and relieving the player from her current primarily forced defensive positioning.

The player often plays from the baseline which results in her often horizontal movement, whilst she opts to use her vertical movement when looking to take a more offensive stance. A result of the player's willingness to move vertically and into the net often leavers her susceptible to attacking play from her opponent. Yet, when the player opts to come to the net as a result of an attacking play, she tends to have more success opposed to simply positioning herself at the net to endeavour to take the upper hand. Amiss from the player's game is the use of angular positioning. A consequence of the player's scarcely integrated use of angular positioning sees her on the defensive quite often on the baseline and an underlying reason why she opts to get to the net as quick as possible. Yet, with the player often approaching the net to then have an error forced, she is more suited to time her approach with the integration of angular positioning to place herself in a more offensive position. If the player can look towards evolving her game to use angular movement, it will not only encourage her to extend through her shots, but place her in a better position to react to the oncoming ball as she learns to position her hips in relationship to the ball and court.

(Ranking, Pre: 73) Ekaterina Alexandrova

When serving the player activates her lower body, yet accentuates the drop in her outside hip and elbow before extending towards the ball. Although the player generates upwards momentum, a result of her accentuated drop on her outside restricts her forwards momentum as her inside falls back slightly more than optimal. Although the player exhibits a sound technique, this slight increase in external drop limits the player's forwards momentum. A consequence of the player's slightly off-cue technique is the speedier upper body – as the player swings through impact her upper and lower body becomes out of sync. Although only slight and not necessarily consistently apparent, this level of play comes down to the slight intricacies of play more readily available to pinpoint when specific technical parameters are key. A slight correction in the player's outer drop and a ball toss more inside the court will correct this behaviour and provide the player with a more fluid and effective serve.

The player's groundstrokes exhibit an activated lower body, coupled with an optimal height of point of contact as the player leans into the court – an optimal performance parameter. Although limited in her momentum into the court, the player's technique enables her to keep on the offensive side of play. Further, both her backhand and forehand see the player's upper body rotate back before transitioning through the ball, although the player does tend to cut off her extension short, pulling up early, or wrapping around – as with her backhand. Yet, the player does apply optimal performance parameters, her angular positioning is a slight limitation and she tends to stand front on at times opposed to repositioning her stance to better allow herself to extend through the ball. The integration of angular positioning into her groundstrokes would largely address the player's current limitations as well as encouraging the player to extend through her shots opposed to pulling back and giving up at times her offensive position.

When returning serve the player activates her lower body as she rotates her upper body, however her height of point of contact is higher than optimal resulting in the player's lower body near extending and unable to transition through her return. The player's tendency to lift post contact limits the player's forwards momentum through the ball – post contact the player remains in a near identical position as pre-contact, whilst her body weight transfers either down, or backwards. The player primarily maintains her positioning, limited in her adjustments pre-contact that result in the player impacting near front-on opposed to the use of angular movement. This combination of limited movement through her return, as with her transition, halts the player's effectiveness and frequently keeps the player on the defensive.

The player moves horizontally more often than not with limited use of vertical movement. Further, the player's ability to apply angular movement through her game is scarce and often where the player's positioning is a weakness. Integration of angular positioning in her game, as with a more consistent activated lower body, are key to the player progressing. Most significantly, the player's control of her height of point of contact is a primary performance indicator that is in need of consistency through play.

(Ranking, Pre: 74) Samantha Stosur

Keeping a low centre of gravity when returning serve, the player exhibits an upper body lean into the ball whilst transitioning through. Although limited in her extension forwards and through the ball, the presence of dynamic balance keeps the player with a maintained activated base for the next ball. Come contact, the player tends to remove her lean into the ball and has a more upwards stance, hindering her release of kinetic through her return. At times on her backhand the player tends to lift, meeting the ball outside of her optimal point of contact. This inconsistency with height of point of contact takes a step away from the elite *elites* and to regain a more optimal return this is a primary performance indicator to be controlled.

With lower body flexion present in her serve that generates an upwards drive, the player at times exhibits optimal performance indicators. The player manages to rotate her upper body whilst lifting her elbow up and through the ball with a more often than not use of the kick serve. Courtesy of her lower body the player is able to push upwards to the ball, yet her momentum post impact is limited, tied to the position of her ball and her racket faces upwards and through motion, contrast to a more distinct through and down action – courtesy of the varying service techniques in the game. Yet, when the player opts to close her racket face and move her ball toss slightly inside the court, her technical parameters near the elite *elites*. Progress on consistency of the player's push through her upper body and momentum post impact will see her achieve a more optimal service motion.

The player maintains an activated lower body through her groundstrokes, yet is limited in the depth achieved more often than not. With torque achieved through her forehand, whilst her backhand tends to maintain its depth, the player does exhibit optimal performance indicators through her groundstrokes, merely limited in her consistency. A primary indicator that halts the player's level of performance is her limitation in applied angular movement, keeping her baseline positioning more frequently. Further, the player tends to hold off on her release post contact on her groundstrokes to naturally pull her forward that is similarly affected by the player's closeness to the ball – catching the ball slightly late at times that restricts both her ability to extend, and fall into the court post impact.

The player does tend to prefer horizontal movement, although isn't shy to move vertically when the ball falls short. While angular movement is readily achieved with the player exhibiting good rotation through her hips coupled with the player's natural release of her torque, the player holds back on her ability to step into the court and cut off the angle of play. Therefore, whilst the player is a good mover, she does get caught catching the ball slightly late, a result of her preparedness and evident proximity to the ball. Progress in these primary areas will enable the player to more frequently exhibit optimal performance parameters and build her game further.

(Ranking, Pre: 75) Yoshihito Nishioka

When serving the player begins with an activated lower body, however is limited in his use of kinetic energy. Restricted with his upwards drive to the ball, the player has a more narrow service motion that halts his momentum, further attributed to the location of his ball toss. A result of the player limited in his lift up to the ball causes pressure to be placed on the player's upper body – although rotation is present, this combination results in the player's push through the ball with limited momentum, as with the player at times wrapping around his serve that limits his ability to openly transition through his serve and use elastic and kinetic energy to his advantage.

The player flexes at the hips prior to returning with limited lower body activation, whereas although this range of movement is deepened, at contact the player more often lifts with limited momentum behind the ball. Whilst upper body rotation is present, the player's height of point of contact varies that results in the player having varied post-contact swing patterns in contrast to a smooth extension through the return. At contact the player is limited in his ability to move with his return, more often back-stepping behind the baseline to set for the next ball – placing himself in a near immediate defensive position.

When contacting his groundstrokes the player flexes his lower body more often on his backhand with a tendency to lift through his forehand. Whilst the player does rotate through his shots, the proximity to the ball is often overstepped with the player caught too close to the ball which forces him to redirect the pressure to his upper body, restricting the use of his kinetic energy. Yet, the player from time will produce a forehand that begins to exhibit optimal performance indicators, whilst his backhand stands quite tall from the hips and the player is limited in his ability to extend through the ball. Primarily remaining on the baseline with restricted momentum behind his groundstrokes, to progress the player is in need of progressing his ability to extend through the ball, coupled with a more consistent lowered centre of gravity. Further, the player is in need of adjusting his proximity to the ball more readily to allow him to move with the ball opposed to getting caught in an often closed position.

More often than not the player is based behind the baseline with limited use of vertical movement. With a favoured horizontal position, the player moves sideways effectively, yet is continuously pushed back behind the baseline as a result of his restricted use of momentum, coupled with his intrinsic recovery position. The use of angular movement is scarce whereas the purposeful application of angular movement would see the player step up from the baseline, whilst also affording the player more time to adjust his proximity to the ball and begin to extend through his shots.

(Ranking, Pre: 77) Taro Daniel

A service motion that is technically sound, from an activated lower body that results in forwards momentum into the court, to a rotated upper body that hits through the ball come contact. The player is slightly limited in his overall upper body rotation, holding back on the speed of his rotation through the ball. This is further attributed to the limited torque achieved that would progress with slight flexion at the hips and contribute to his ultimate upwards drive up and out into the ball.

Not afraid to step into his return on his backhand, the player does tend to have a higher point of contact on his forehand, yet does manage to use his momentum against the oncoming ball. The player has a tendency to move sideways with the ball upon return that sees his upper body take over. Maintaining a low base of support for the most part, the player moves angularly at times that sees him cut off the server's angle and take an offensive stance. The player is slightly limited here in his ability to hit through the return, pulling back at times, hindering the effectiveness of his return and post-impact positioning.

The player positions himself with a wide base of support when performing his groundstrokes whilst maintaining flexion at his hips in a stance that allows him to move through the ball. On his backhand side the player tends to slightly push through his upper body when unable to achieve optimal positioning, yet on his forehand tends to jump through if not in his optimal range of impact, as the player uses momentum to his advantage. The player's strength is in maintaining a low centre of gravity through impact, an optimal performance parameter, that allows the player to transition through his shots, consistently. A noted limitation is the player's consistency in setting for the location of the ball – a slight further extension on his forehand side, and slight adjustment to his proximity to the ball on his backhand, would progress his groundstrokes towards the elite *elites*.

The player utilises his movement to near optimal performance. Not afraid to move angularly at times, the player applies horizontal and vertical movement as necessary that are optimised through his consistent lowered centre of gravity. The use of dynamic balance throughout allows the player to transition through his shots, whether vertically into the court, or angularly post impact upon anticipation to the next ball. The player's weakness is largely in his location of impact – his proximity to the ball to ensure he hits through opposed to finishing too early.

(Ranking, Pre: 79) Tamara Zidansek

Upon return the player keeps a wide base of support, lowering her centre of gravity through contact with a more often than not optimal height of point of contact. As the player rotates through her upper body, at times she catches the ball too close, restricting the player's ability to hit through the ball. Whilst the player does appear to more often transition through her returns, she is limited in her forwards lean and resultant momentum behind the ball. Yet, from time the player does exhibit optimal performance indicators, her consistency in their application is often hindered by the player's near stationary position through, and at times, post contact.

When serving the player maintains an activated lower body coupled with flexion at her hips that pushes her into the court – using her kinetic energy behind the ball. The player equally manages to rotate through her upper body and release her elastic energy through the ball, with a slight limitation evident in her upwards drive as she begins to release towards the ball. A primarily technically sound serve, the player places her ball toss inside the court and achieves momentum behind the ball. At times the player does finish with a taller upper body, indicative of her hold-back on her momentum. A more open release through the ball will work with the player to progress her service motion.

Through her groundstrokes the player often tends to keep a lowered base of support to meet the ball at the optimal height of point of contact. However, the player does tend to catch the ball slightly higher than most optimal, limiting her ability to transition and extend through her shots. Whilst optimal performance indicators are apparent through the player's forehand and backhand, her level of consistency is key. The player scarcely transitions through her shots as well as often being caught on the defensive behind the baseline. Yet, when the player takes an offensive position, her optimal performance indicators are more consistent. The player is restricted in her ability to extend through her groundstrokes, whether on the offensive or not – largely a result of the player staying tall through her upper body.

The player primarily maintains a baseline position that results in her horizontal movement more often, with vertical movement a result of the player moving back as she endeavours to control her height of point of contact. More often, however, as the player moves back she does get stuck on the defensive with scarcely applied angular positioning or extension through her shots to regain ground. A consequence of the player's often backwards movement, or horizontal play, is her limitation is consistently stepping up. Whilst the player can and does from time step into the court, more often she remains on the baseline. To progress, the integration of angular positioning will not only promote extension through the player's shots, but primarily place her in a position to maintain an offensive stance and take time away from her opponents.

(Ranking, Pre: 85) Polona Hercog

When serving the player begins with a step-up approach to help her gain momentum behind her serve whilst keeping her lower body activated. As the player extends her upper body she generates a push through her legs that allows her to generate an increased release of her kinetic energy through the ball. Whilst the technical parameters of the player's serve are near optimal with her momentum into the court, she does tend to keep her racket face dropped before she opts to reach and extend – this slight pause in her serve fluidity takes time away from the height of her ball toss – as with the elastic energy generated through her upper body. A slight adjustment to the wait here, to a smoother transition, would enable the player to use here elastic energy more so to her advantage.

Upon return the player keeps her lower body slightly flexed as she leans into the court with her upper body. As the player manages to keep a controlled height of point of contact, although inconsistent, her proximity to the ball is her greatest challenge – a result of her anticipation. Whilst the player has evident technical indicators to excel through her returns, she is held back in her inconsistency as with her ability to apply angular positioning. To progress, timeliness is key for the player to set herself prior to impact more readily, whilst working towards the location of impact – at times slightly too close, or in a position to extend, yet her feet do not follow. Coupled with a more pronounced level of flexion in the player's lower body through impact will see the player begin to harness a more effective return.

When contacting her groundstrokes the player has a quite pronounced loop on her forehand that sees her elbow lift alongside her racket face, reaching a near vertical point, opposed to a smooth circular motion. This varied technique results in the player mistiming her ball from time, as with her proximity to the ball, restricting the player from extending through her forehand – often caught in the same position post-impact. At times the player will rotate her hips through the ball, although this is a scarce performance indicator. On the player's backhand she is more technically sound with a fluid motion – whilst although she does have a tendency to pull up through the shot, at times she will lean into the ball – yet an inconsistent part of her game.

The player's movement with the ball is perhaps her greatest weakness. Although a relatively smooth mover on court, with horizontal movement preferred, the player's awareness of her positioning and level of anticipation is often cause for the player to meet the ball closer to her body than most optimal, or in a stance that limits her ability to extend through her shot. With a more consistent lowered centre of gravity, as with a more consistently controlled height of point of contact, the player can begin to integrate a smoother transition between shots – often rushing through her groundstrokes. As the player allows herself more time to set before her shots, with a current tendency to rush, and with a more purposeful use of angular movement, she will begin to progress.

(Ranking, Pre: 87) Eugenie Bouchard

When returning serve the player flexes as the hips, leaning her upper body into the return as she activates her lower body before transitioning through the return. When given the opportunity, the player steps up through impact to transfer her kinetic energy through the ball, lifting her body forwards to counteract the force of the ball. With good upper body rotation through impact, the player is slightly limited in her lower body flexion at times that affects her height of point of contact. Whilst exhibiting the characteristics of optimal knee bend and height of point of contact through impact, consistency is key. The player's weakness is in her anticipation of the ball that also affects her proximity to the ball at impact and ability to extend through the ball.

As the player serves she achieves optimal lower body activation she uses to drive herself upwards towards the ball. Further, the player exhibits near optimal extension and momentum behind her serve – an optimal performance indicator. Whilst the player rotates her upper body she achieves a smooth elbow drop to generate elastic energy up and through impact. The player is limited in her torque that would enable her to further release through the ball at impact. Further, the player's ball toss is slightly closer than most optimal – moving this out slightly alongside an increased push through the legs would progress the player's serve towards the elite niche.

Through her groundstrokes the player not only exhibits optimal performance parameters, but matches the performance indicators of the elite with a deep level of flexion through impact off both sides, as with her height of point of contact. Whilst the player rotates her upper body through her groundstrokes, she opens her hips on her forehand to use torque to her advantage – an optimal performance indicator. It is no surprise that this player was once a part of the elite niche, whereas her primary weakness is in her anticipation of the ball and her consistency.

The player is primarily a baseline player that moves horizontally most frequently, yet does make minor moves vertically – up or back – as she endeavours to meet her groundstrokes inside the optimal range of impact. The player quite frequently exhibits characteristics of the elite *elites*, however her current ranking is far from this niche. This is a consequence of the player's limited integration of angular movement, although present at times, is yet to be used point by point. However, seemingly the player's most significant weakness is in fact her timing which affects her proximity to the ball, her level of lower body flexion, as it remains inconsistent, plus the deeper integration of angular movement. Although slightly held back in a number of areas to do with the player's consistency of application, a more streamlined game will see the player prosper again as she once did.

(Ranking, Pre: 90) Cameron Norrie

The return of the player begins with his stance slightly higher off the ground than most optimal. This initial stance hinders his point of contact height and the player's ability to utilise his stored elastic energy. From the beginning of the return, perhaps without knowing, this player puts himself on the back foot. A beneficial technical aspect is his use of momentum – not shy to move with the ball upon impact at times, although not consistent.

The service motion has lower body flexion and solid upper body rotation. Using a step-up stance at times limits his lower body flexion with the urgency to get into a set position before lifting off. The location of his ball toss could be slightly more in front to encourage the slightest of 'added' reach and thus momentum to move into the court. Although recovery post serve doesn't seem to be a problem, there is room here for the service action to move to the next level.

When it comes to his groundstrokes the player sets nicely on his forehand side, however is limited in his post contact movement. The same applies for his lower body – preparation is solid, yet as contact lingers as he begins to lift prior to contact, disposing of his pre-stored energy whilst raising his height of impact unnecessarily. The backhand, however, is reached into nicely and the player's lower body is set and used to its advantage. It becomes apparent here that horizontal movement is a strength, yet when a vertically inclined directional change is key the player is mostly stuck at contact which limits his ability to maintain an offensive rally stance, placing him more often on the defensive; a consequence of the player being scarce in his applied angular directions.

(Ranking, Pre: 92) Denis Istomin

When serving the player uses his levers to his advantage as his upper body extends towards the ball. With the player activating his lower body more often than not, at times exhibiting only slight flexion, the player pushes himself upwards to hit out through the ball – an optimal performance indicator. As the player flexes through his serve, his hips at times tend to sit even – in a horizontal position, contrast to the in-court lean required to position his centre of gravity inside the court. Yet, the player has near optimal technical parameters through his serve and is merely limited at times by his in-court transition and location of contact that sees his ball land slightly short in the service box.

Upon returning serve the player lowers his centre of gravity towards the ground and keeps a lowered base of support throughout, accompanied by slight flexion at the hips seeing the player lean ever so slightly into his returns. Although inconsistent as the player tends to sit slightly taller at times through his return, when lowered he has a more often controlled height of point of contact off both sides. Yet, the player is limited in his ability to extend through his returns as he also opts to maintain his position through impact with limited use of momentum behind the ball. Further, the player rotates his upper body as he returns, yet tends to catch the ball slightly late that limits his ability to extend through the return.

The player maintains flexion in his lower body when contacting his groundstrokes, whilst his height of point of contact is primarily controlled. At times the player's level of flexion is slight which results in a slightly higher point of contact, however the player does tend to exhibit optimal performance parameters as he varies in his consistency of use. On the player's backhand, however, he does often pull up short and have his upper body quite tall, contributing to his inability to transition and extend through the shot. Yet, the player's forehand is slightly more consistent as he is more comfortable leaning into the court as he hits through the ball and at times, allows his momentum to bring him further inside the court. Whilst the player does exhibit optimal performance indicators he does need to flex his hips more consistently to help encourage that forwards lean to progress towards a more consistent transition and extension through the ball post impact.

More often than not the player moves horizontally along the baseline. At times the player shows glimpses of his use of angular positioning as he rotates through his hips, however the player is limited in its use. Whilst vertical movement is used more often to recover after he transitions into the court, the player isn't shy of moving forward where he's able to use his longer levers to his advantage. Yet, the scarce use of angular positioning keeps the player more often on the defensive and unable to take time away from his opponents, not gaining ground and subsequently taking an offensive stance through play. To step up to the elite, the player is in need of a more consistent use of angular positioning and his ability to extend through his balls, using momentum to his advantage.

(Ranking, Pre: 97) Katie Boulter

The groundstrokes of this player are complemented by the use of her lower body. Maintaining a solid state of balance throughout her rallies, it becomes apparent that her weaker area is rather in her control of the ball – hitting too far through her shots.

Positionally, the player can move both horizontally and vertically quite well with speed and agility on her side. Meanwhile, angular positioning is rare and a key performance indicator the player is in need of to further progress. Similarly, the player's ability to read play, or where the ball is heading still needs development – her responsiveness to play, or readability, is a weaker area. This limitation stops the player in her tracks mid-rally when the ball is returned either behind her or where she would have been standing if her reaction was in sync with her opponents' racket trajectory upon contact.

When returning serve the player's height of impact is higher than optimal as is her limited lower body flexion. Her movement into the return isn't a consistent part of her game, although it's clear she can and does, simply not as often as demanded.

When serving the player's upper body rotation is limited, compounded by the equally limited torque throughout her service motion. Stemming from her restricted lower body drive, her upwards lift is in need of further development. The presence of a controlled centre of gravity allows the player to maintain a sound service motion, yet it's effectiveness in putting the player immediately on the offensive each respective service game is less than desirable. All round development is key for this player to make a more pronounced step in the rankings and to toe the line with the elite niche.

(Ranking, Pre: 99) Reilly Opelka

The player exhibits an activated lower body when returning serve, however is limited in his use of momentum behind the return. With flexion present at his hips, the player does lean into his return without lifting. At contact the player tends to hold back on his transition through the ball, with limited rotation coupled with a restrained follow through more often. The player appears to make contact with the ball closer to his body than most optimal – the player's proximity to the ball is in need of further adjustments to progress the quality and consistency of his serve return.

A service motion with an activated lower body, the player generates deep flexion that identifies with an optimal performance indicator. With upper body rotation present as with the release of elastic energy, the player is capable of applying significant momentum behind his serve – yet holds back at times. It becomes apparent that although the player has longer levers he uses to his advantage, the upwards lift generated is stalled as a result of his ball toss height, and location. Whilst more often identifying with optimal performance indicators, two key areas are the player's closed momentum, as he holds back and stops short during his follow through, alongside his reach up, through, then down – through his serve with an apparent restriction in the fluidity of the player's service technique.

When contacting his groundstrokes the player often keeps an activated lower body off both his forehand and backhand sides. With upper body rotation evident through the player's follow through, his weakness is clear in his proximity to the ball as he is restricted in his ability to extend – pulling off early opposed to committing to his shot. Further, the player's follow through tied to his rotation, at times is hindered, compensating for his open racket face upon contact at time on his forehand. Off the player's backhand he has a tendency to fall back, still in need of progression with his moveability, extending through his shots, and ability to adjust before impact to limit his tendency to overstep his location of contact.

Whilst the player has longer levers to use at his advantage, this also results in the player's limitation in his fluidity, still adjusting to his proximity of the ball throughout his game. Whilst limited in his vertical movement, as well as angular positioning, horizontal is the movement of ease and applied. The player's disadvantage is in his limitation to apply momentum which results in the player keeping a defensive stance more often than not. Progression with the player's angular movement, as with his ability to extend through his shots, coupled with his adjustments to the location of impact, will allow the player to move more openly and begin to benefit greatly from his longer levers and move deeper in the rankings.

(Ranking, Pre: 125) David Ferrer

Lower body propulsion, forwards momentum, torque coupled with consistent shoulder rotation provides the quality service motion of a seasoned tour player. The player allows his service momentum to push him into the court whilst his 'bounce back' step allows a quick recovery to stand firm on the baseline.

Although not the strongest or tallest player on tour, the player is reliant on his speed of movement and consistent play. It is of no surprise then that the return of serve isn't his strongest point when the opponent has a heavier and powerful serve to return. He does well, however, by utilising his lower body mechanics to his advantage, yet his prone defensive play over an offensive stance leaves openings for his opponent.

The player's footwork solidifies him as a worrisome opponent no matter who he comes up against – after all, he is a former top tier player. With horizontal and vertical movement coming with ease, angular movement and its depth of integration is not as significant as it once was – or is necessary to claim a spot inside the elite niche, again. Indeed, the results of the past have lessened mostly due to his stature and power prone weapons of his opponents, regardless of the player's speed and agility, tied to his less pronounced use of angular play.

When considering his groundstrokes, he is easily more comfortable on his backhand due to the added power provided by the extra hand, favoured by most player's all round. The rotation on his secondary side and open stance is a consistent weapon of choice. Likewise, the player's forehand is solid, yet rotation isn't as present – an area that although minor, hinders the player's forehand from causing trouble for the more elite status opponents.

(Ranking, Pre: 239) Jo Wilfred Tsonga

The player's return height of contact at times is higher than most optimal, a result of remaining stationary prior to contact opposed to moving upwards, or slightly back, to better gauge and prepare for the height of the oncoming ball. The player steps back on his backhand side come contact, whilst forwards on his forehand when opting to re-position himself prior to impact, at times. This inconsistency at contact results in varied heights of contact whilst also putting the location of the ball at contact in fluctuating positions, limiting the player's ability to extend through his shot openly.

Exhibiting smooth momentum on his serve, the player is equally cautious on this forwards transition when his opponent has taken the upper hand. Whilst lower body activation and the incorporation of hip flexion is apparent, upper body rotation is limited at times that sees the player restricted in releasing his elastic energy through contact. The player drives through his lower body, yet his upwards lift is limited, tied to his ball toss not as out in front as necessary to keep his momentum moving in his favour.

A strong player with immense power generated off the ground through his groundstrokes, his upper body generates a wide range of rotation that the player then uses to unleash into his groundstrokes. The player's preparedness is key as he has ample time for a pause prior to impact – an optimal performance indicator. Further, the player's use of timeliness, as he manages to better control his centre of gravity and maintain a balanced state, projects through his shots on both sides. The slight weakness evident in the player's groundstrokes are tied to his varying height of contact when more pressure is placed onto his upper body, which is often the result of maintaining a horizontal position opposed to moving vertically and/or angularly into the court.

When considering the player's movement, his preparedness is a strength which indicates his agility and shot anticipation as those of the elite *elites*. Although slight, the player does favour horizontal movement, whereas stepping into the court more often and taking advantage of his already sound angular movement to take time away from his opponent, coupled with that a more offensive stance, would see the player progress even further, as once before as a former elite *elites*. It is key to note that when the player opts to use his momentum and step up into the court he is incredibly dangerous with more often positive results, identifying as the calibre of player he was, and has the capacity to be again.

(Ranking, Pre: 240) Andy Murray

When serving the player exhibits deep lower body flexion that is used to propel him upwards towards his ball toss – an optimal performance indicator. The player rotates his upper body as he also achieves torque courtesy of the rotation through his hips. At impact the player moves into the court with momentum on his side – a quality primarily of the elite *elites*, often held back on by lower ranked player's, indicative of his current ranking. Not only does the player exhibit technical parameters of the elite *elites*, he maintains his ability to consistently deliver on optimal performance parameters.

Upon return of serve the player widens his base of support before stepping up to meet the serve at the optimal height of point of contact – an optimal performance indicator. The player also abbreviates his swing as he returns, unless left with ample time before impact to allow him to set and start off on the offensive. The player's upper body rotation, as with the torque consistently generated through his shots, is followed up by the player's momentum behind the ball – another optimal performance indicator. With the player's lowered base of support, optimal height of point of contact and ability to extend through his returns, he exhibits optimal performance parameters that match his ranking 'prior' to his current injury-induced time away from tour that has resulted, primarily, in his ranking slide.

The player maintains deep flexion in his lower body through contact on both his forehand and backhand that allows him to maintain an optimal lowered height of point of contact – exhibiting optimal performance parameters of the elite *elites*. The player's upper body rotates openly through his groundstrokes, although he is at times limited in his extension through the ball – an inconsistent aspect of his current game. At times the player does tend to close off on his backhand in contrast to extending through the ball which is a weakness as he pulls back and typically ends in error. However, primarily the player is consistent off both sides and maintains an offensive stance more often than not, further courtesy of the player's ability to use his torque to his advantage. The player does tend to hold back off his momentum into the court at times, tied to his limited extension through the ball. Yet, more often than not the player does showcase qualities of the elite *elites* and performs likewise.

With the integration of angular movement through play, the player can move both horizontally and vertically with ease, although his angular positioning is a significant strength – an optimal performance indicator. Yet, a result at times of the player pulling back opposed to transitioning through, places a heavier reliance on his hips and the torque achieved – interestingly a significant causation of his ongoing injuries. An open release of the player's hips followed by the player's lean into the court, will begin to alleviate the pressure off his hips. If the player can build on his transition through his shots in contrast to holding steady and releasing through his hips, it will equally take the load off his mid-section whilst transitioning through his shots.

Part 2: Conclusion

The technical properties of The V by Dr. B © have been unleashed. Their real-world *transference* has been disclosed. Methodically, the *one-percenters* maintain their advantage. Those who exhibit cravats in their application of the criteria and its reciprocity through transference lose their position against a *one-percenter* or, as a elite *elites*. Meticulously, the higher the consistency in application delivers greater *all-round* transference – awareness is key.

Fluidity in motion is the *end-point* of functional technical parameters. The naked eye views this with ease, yet in *reality* thousands of hours go into this streamlined process. Quality in repetition builds the autonomy loop. The V by Dr. B © *feeds* the blueprint of optimal performance parameters and *sets* the fundamental performance foundations to develop into a *one-percenter*.

The *fore*hand groundstroke sets the scene. Player awareness of its properties and the criteria of The V by Dr. B © results in a *one-percenter*. Lack of concept or relational awareness is a consequence of a weaker ranking. Likewise, inability to apply transference, as indicative of *angular* integrations, contributes significantly to ranking failure. Reciprocity and meticulousness are prime attributes of the *one-percenters*.

Whilst in conjunction with the *Optimal Performance Theory* (Berge, 2017), The V by Dr. B © guarantees an outcome as a *one-percenter*. As absurd as this historical implication reads, its disclosure is the mere *reality* and *truthfulness* of science – its capacity when relayed to the sporting world. Likewise, the *ninety-nine* elite players measured against its principles, in accordance with The Rule of Transference, were the *first* of their calibre to have their performance as a *whole* weighed against its respective properties – the criteria of The V by Dr. B ©.

The break-down is transparent – technical parameters rein in conjunction with player fluidity, or rather, the implications of functional performance parameters. Limited extension, or a pull-back to prevent an open transition is a primary example of a restrained performance halting a player's release through impact – correlated with *point of contact.* Likewise, restricted range of movement or reservation in lowering a player's centre of gravity increases pressure on the upper body, whilst delimiting the body's natural acquirement of kinetic energy – correlated with *knee bend.* A player's moveability is dependent on their transition between horizontal *and* vertical movement, yet to account for the sequence of *discrete skills* for both *point of contact* and *knee bend*, angular movement *and* positioning is a key facet of optimisation – correlated with *balance*.

Recall, losing the ability to perform at the frequency upheld by The V by Dr. B © is denoted by failed results as a *one-percenter* – a commonality on *both* tours whilst a select few manage to *maintain* their ranking hold. The *three* prime examples, respectively, at time of publication, include: Novak Djokovic, Roger Federer, Rafael Nadal, Simona Halep, Angelique Kerber and Naomi Osaka. These players have not only maintained a consistent *one-percenter* status, but have kept their advantage. The remaining *one-percenters* often have a limit – a higher tendency of ranking failure. Interestingly, these players are *not* necessarily representational of the *top three* on their respective tour. A consequence of ranking fluctuations – those who *maintain* their hold without succumbing to failure are players *who* showcase the criteria of The V by Dr. B © in action more frequently *through* action. As disclosed through *Coaching Analyses* this is not a coincidence, rather a direct correlation now disclosed.

PART 3

THE STATISTICS

Part 3: Introduction

To complement the aforementioned disclosure, the *truth* in numbers further solidifies the historical implications of The V by Dr. B © and its pathway to an elite *elites*. Never before has a trail been afforded synonymous with all too often the objective – *top 10 in the world.*

The Unveiling disclosed The V by Dr. B © and its depth of analysis – simplified. Whilst simplification may not have been apparent, *The Statistics* signifies its *complementary* half. *Coaching Analyses* placed the *truth* in *The Unveiling* into the *real-world* before unveiling *The Statistics* herein with its statistical inferences.

Rudimentary listed as profiled in *The Unveiling*, each variable is further disclosed, individually and mixed in alignment with the now familiar criteria. Methodically, variations of the mean, mode, frequency, standard deviation (*SD*), range, cumulative percent and Pearson's correlation are *released*, formally, and their statistical significance (2-tailed).

Fair warning, *The Statistics* are in correspondence to *The Unveiling* and for the statistically inclined. *The Statistics* recalls *The Unveiling* in-addition to their statistical inferences with *less* apparent data sets and the corresponding variables. Recall, reference to *The Statistics* throughout *The Unveiling* for a more pronounced look into the numerical values.

A consequence of its statistical insights, a quantitative collection of the *in-conjunction to* descriptive statistics remain. The list is in accordance with its counterpart, *The Unveiling*, for ease of reference whilst *scientifically* further establishing the *proven* technical parameter – The V by Dr. B © – whilst accounting for *all* variables, singular and mixed, and their statistical inferences. *The Statistics* incorporates prior reported variables elaborated, and newly presented variables that were withheld from *The Unveiling* as a consequence of their scarce or limited statistical significance.

To preserve the sequence of *The Unveiling*, the list follows suit. For this reason, the aforementioned variables and mixed criteria are recommended to be viewed *in-conjunction with* their further disclosed statistics, herein, for greater depth of association.

No knee bend

The mean was found to be 0.0 ($SD = 0.0$), no knee bend was recorded on the Observation Chart out of the total 123 participants observed. The frequency of no knee bend I reported, that is 0.00, is 123.0 and the cumulative percent is 100.0.

The histogram illustrates that no knee bend occurred out of all participants observed performing the forehand groundstroke at point of contact.

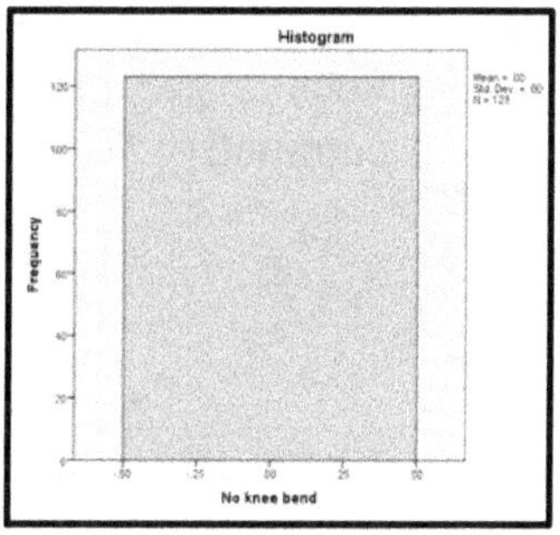

Figure 30 Statistical analysis of no knee bend in the 123 players observed

Slight knee bend

The mean is 7.3 and the mode 3.0 (*SD* = 5.0) and the range 24.0. The frequency of slight knee bend is >5 but <10 at frequencies: 1, 3, 4, 5, 7 and 8, the cumulative percent is 100.0.

Deep knee bend

The mean is 22.7 and the mode 26.0 (the lowest out of multiple modes) (*SD* = 4.9) and the range 24.0. The frequency of deep knee bend is >5 but <10 at frequencies: 22, 23, 25, 26, 27, 28 and 29, the cumulative percent is 100.0.

Balanced

The mean is 28.8 and the mode 30.0 (*SD* = 1.3) and the range 8.0. The frequency of balance is >23 but <36 at frequencies: 29 and 30, the cumulative percent is 100.0.

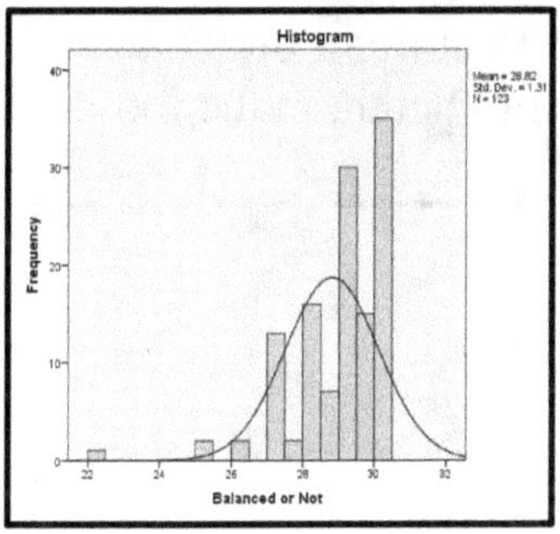

Figure 31 Statistical analysis of balance or not in the 123 players observed

Contact above

The mean is 3.0 and the mode 1.0 (*SD* = 2.2) and the range 12.0. The frequency of contact above hip is >10 but <20 at frequencies: 1, 3, 4 and 5, the cumulative percent is 100.0.

Contact hip

The mean is 7.3 (*SD* = 3.1) and the range 20.0. The frequency of contact at hip is >11 but <16 at frequencies: 6 and 7, the cumulative percent is 100.0.

Contact below

The mean is 19.6 (*SD* = 4.0) and the range 22.0. The frequency of contact below hip is >7 but <11 at frequencies: 17, 18, 19, 22 and 23, the cumulative percent is 100.0.

Slight/hip

The mean is 2.5 (*SD* = 2.4) and the range 12.0. The frequency of slight knee bend and contact at hip height is >10 but <15 at frequencies: 1, 2 and 3, the cumulative percent is 100.0.

Slight/below

The mean is 2.0 (*SD* = 3.3) and the range 17.5. The frequency of slight knee bend and contact below hip height is >14 but <18 at frequencies: 1 and 2, the cumulative percent is 100.0. The histogram illustrates the frequency of 1 as the most commonly occurring number (zero excluded) out of the 123 participants observed at point of contact with slight knee bend and contact below hip height in the forehand groundstroke.

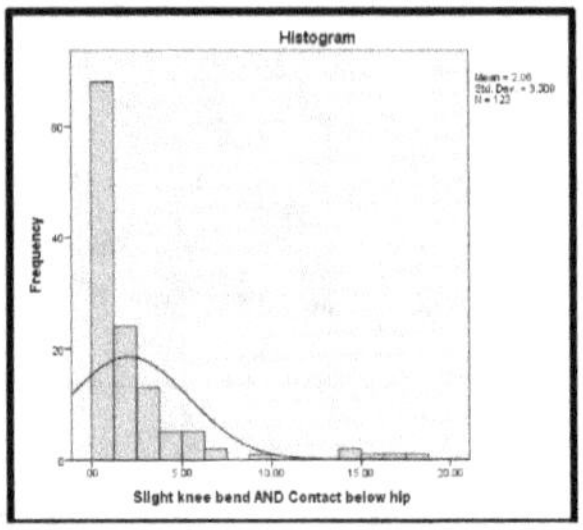

Figure 32 Statistical analysis of slight knee bend and contact below hip height in the 123 players observed

Deep/hip

The mean is 4.7 (*SD* = 3.1) and the range 14.0. The frequency of deep knee bend and contact at hip height is >10 but <15 at frequencies: 2 and 4, the cumulative percent is 100.0. The histogram illustrates the frequencies of 2, 3 and 4 as the most common occurring numbers—however due to the curve, frequencies 6 and 8 appear lower although have a higher frequency than the number 3 (9 over 8).

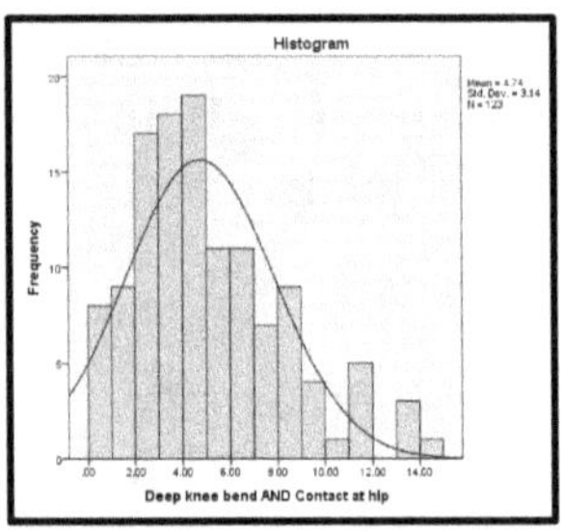

Figure 33 Statistical analysis of deep knee bend and contact at/around hip height in the 123 players observed

Deep knee bend contacting ball below hip height

The mean is 17.5 (*SD* = 4.5) and the range 24.0. The frequency of deep knee bend and contact below hip height is

>9 but <12 at frequencies: 16 and 17, the cumulative percent is 100.0.

Ranking

The mean is 70.7, the mode 6.0 (smallest of multiple modes found) (*SD* = 57.8) and the range 198.0. The frequency of player-rankings is highest at <50, the cumulative percent is 100.0.

Ranking V slight knee bend

When looking at player ranking and the use of slight knee bend during the forehand groundstroke, it was of interest to see if the results would resemble those of Slight Knee Bend. Descriptive statistics were generated for player ranking, that is, the ranking the player held at the time of observation (cumulative ranking mean reported at 70.7), and the number of slight knee bend. The mean was found to be 7.3 (*SD* = 5.0) out of the 123 participants analysed. Player rankings were correlated against the number of slight knee bend in the forehand groundstroke and Pearson's $r = -.158$ with $p = 0.081$ not of statistical significance (2-tailed). Figure 34 illustrates the scatter plot of this relationship.

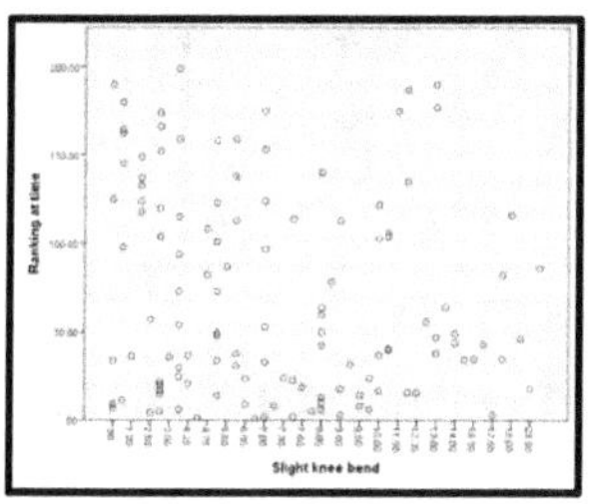

Figure 34 Scatter plot of ranking and slight knee bend in the 123 players observed

The figure shows that slight knee bend is prevalent amongst not only the world's best players, but also among those who hold a ranking outside the top 100.

Ranking V at/around hip height

When looking at the ranking correlation and frequency of height of point of contact at hip height, it was of interest to see if the results demonstrated the frequency of Contact at/around Hip Height when considering the variable singularly. The excluded variable, height of point of contact above hip height (not a characteristic of The V by Dr. B ©), was found to have a mean of 3.0, there are 27 balls out of a possible 30 yet to be defined at height of point of contact. When generating descriptive statistics for player ranking (70.7) and point of contact at hip height, the mean was found to be 7.3 ($SD = 3.1$) out of the 123 participants analysed. Player rankings were correlated against the number of point of contact around/at hip height in the forehand groundstroke and the Pearson's correlation $r = 0.305$ with $p = 0.001$ statistical significance (2-tailed). Figure 35 illustrates the scatter plot, and suggests that players ranked >50 appear to contact on average more frequently at hip height than the higher ranked players (closer to one).

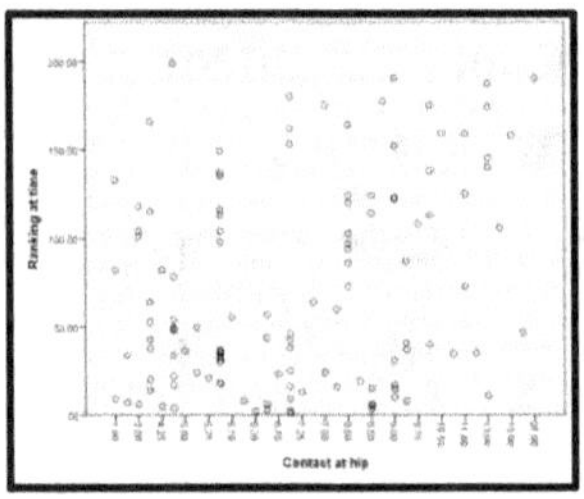

Figure 35 Scatter plot of ranking and contact around/at hip height in the 123 players observed

Ranking V above contact

Descriptive statistics were generated for player ranking and point of contact above hip height. The mean is 3.0. Player rankings were correlated against point of contact above hip in the forehand groundstroke and Pearson's correlation found r = 0.159 for contact above hip with p = 0.079 not of statistical significance ($SD = 2.2$) (2-tailed).

Ranking V slight knee + contact at/around hip height
Descriptive statistics for player ranking and slight knee bend were generated to identify if this consistency (as the second favoured variables) was still present. The mean was found to be 2.5, indicating a small frequency, similar to Slight Knee Bend/Contact at/around Hip Height. The correlation of player rankings against the number of slight knee bend collated with point of contact at hip height in the forehand groundstroke was then considered. Pearson's correlation $r = -0.171$ with $p = 0.059$ not of statistical significance (2-tailed). The scatter plot in Figure 36 illustrates player rankings and the use of slight knee bend with a height of point of contact at/around hip height in the 123 players observed.

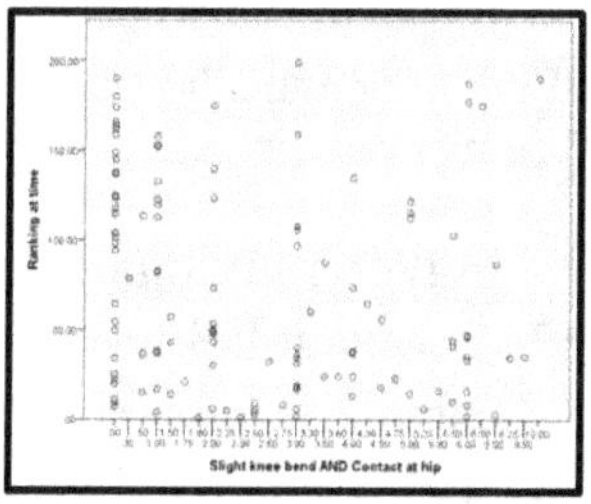

Figure 36 Scatter plot of player rankings with the use of slight knee bend and contact at/around hip height in the 123 players observed

When compared against player rankings, slight knee bend and contact around hip height has a higher frequency on the scatter plot for players ranked inside the top 50. The scatter plot illustrates that players ranked outside the top 50 also apply this combination during the forehand groundstroke, although have a lower frequency.

Ranking V slight knee + contact below hip height
Considering the variables singularly, slight knee bend was not a frequently applied position; height of point of contact below hip height was the most frequent variable. Thus far the mixed criteria, outlined in Mixed Criteria: Knee Bend (deep/slight) and Point of Contact (above/hip/below), were not consistent

without the presence of deep knee bend. The consistency of these results is of interest. Descriptive statistics were generated for player rankings and slight knee bend collated with point of contact below hip height, and the mean was found to be 2.0, illustrated in the scatter plot: Figure 37.

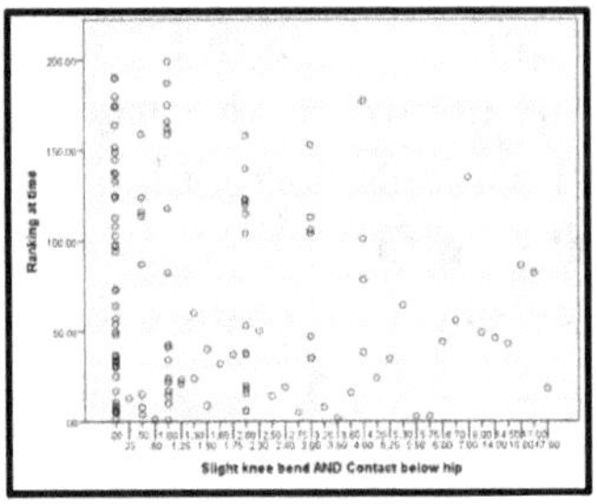

Figure 37 Scatter plot of ranking and slight knee bend with contact below hip height in the 123 players observed

The correlation between player rankings against slight knee bend collated with point of contact below hip height was then considered and Pearson's correlation $r = 0.131$ with $p = 0.149$ not of statistical significance (2-tailed). The minor frequency in the mixed criteria was found to align with Mixed Criteria: Knee Bend (deep/slight) and Point of Contact (above/hip/below).

When compared against player rankings, contact below hip height coinciding with slight knee bend has a higher frequency in the lower echelon on the scatter plot for players ranked inside the top 50. The scatter plot illustrates that players ranked outside the top 50 also use this combination during the forehand groundstroke, although with a reduced frequency.

Ranking V deep knee + contact at/around hip height

When considering the presence of deep knee bend in the mixed criteria, the frequency of results increases (as indicated in Mixed Criteria: Knee Bend [deep/slight] and Point of Contact [above/hip/below]). Whilst generating descriptive statistics for player ranking and deep knee bend collated with point of contact at hip height, it was of interest to identify if an increase in frequency was present, contrast to when slight knee bend was considered. The results found the mean to be 4.7, as illustrated in the scatter plot (Figure 38), aligning with Mixed Criteria: Knee Bend (deep/slight) and Point of Contact (above/hip/below), indicating that the presence of deep knee bend does in fact increase the frequency of results when correlated.

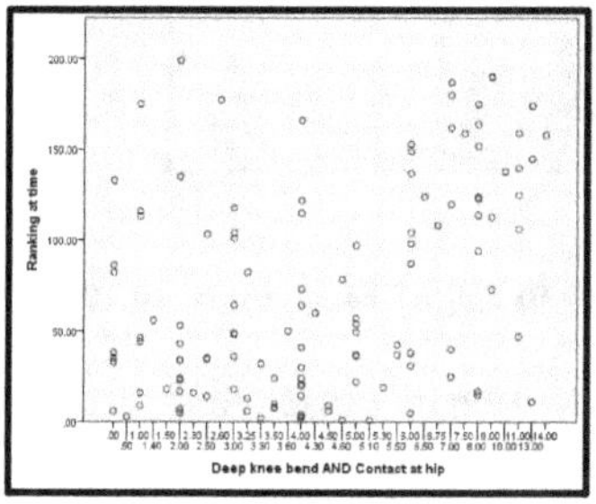

Figure 38 Scatter plot of ranking and deep knee bend with contact at/around hip height in the 123 players observed

In addition, it was of interest to identify if a correlation was present in player rankings when correlated against deep knee bend collated with point of contact at/around hip height in the forehand groundstroke. Pearson's correlation between deep knee bend and point of contact at/around hip height was found to be r = 0.435 with p = 0.000 statistical significance (2-tailed). The results indicate that with the presence of deep knee bend, the frequency increases (4.7), identifying a significant relationship between the two variables. This section also indicates that The V by Dr. B © is applied by the elite when contacting their forehand groundstroke.

Top 10 V slight knee bend

When looking at slight knee bend in the top 10 players, it was of interest to know if the mean was the same as the overall result, or if this ranking bracket differed. Of concern was the variable slight knee bend, and if it occurs less than deep knee bend, as found in Knee Bend. Frequency statistics were generated and the mean was found to be 6.6 ($SD = 4.0$). Interestingly, as the histogram in Figure 39 illustrates, the frequency of slight knee bend in the world's top 10 players is minimal and is lower than the overall mean (7.3).

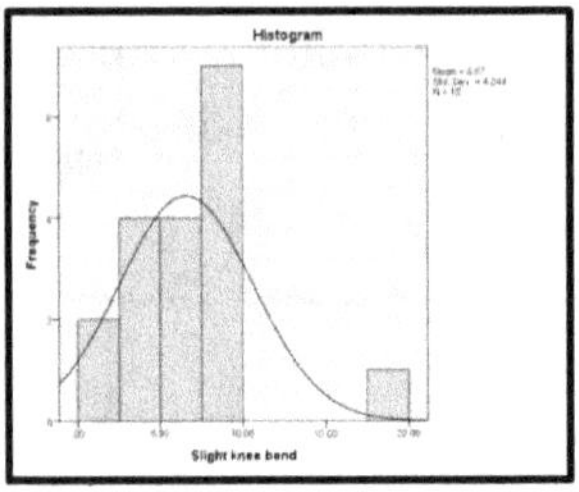

Figure 39 Histogram of slight knee bend in top 10 players

Descriptive statistics were generated for participants with slight knee bend at point of contact in the forehand groundstroke with a ranking inside the top 10 on either the WTA or ATP tours. The total number of participants with a top 10 ranking is 18. The mean is 6.5 ($SD = 4.0$) and the range 17.5. The frequency of slight knee bend reported in top 10 participants occurs twice at 8, all other numbers (zero excluded) fall between >0 but <18 with a single frequency (5.6 % of the cumulative total) the cumulative percent is 100.0.

When compared against player rankings inside the top 10, the table illustrate that players are applying slight knee bend when contacting their forehand groundstroke on average six times out of a possible 30 balls observed. This result indicates that players who hold a top 10 ranking use slight knee bend less than 20 % of the time when contacting their forehand groundstroke.

Top 10 V contact above hip height

When looking at height of point of contact above hip height in the top 10 players, it was of interest to see if this variable was also minimal, due to it not being a characteristic of The V by

Dr. B ©. Frequency statistics were generated and the mean was found to be 1.9 (SD = 1.4). The histogram in Figure 40 also highlights the minimal application of elite players contacting their forehand groundstrokes above hip height, lower than the overall mean (3.0).

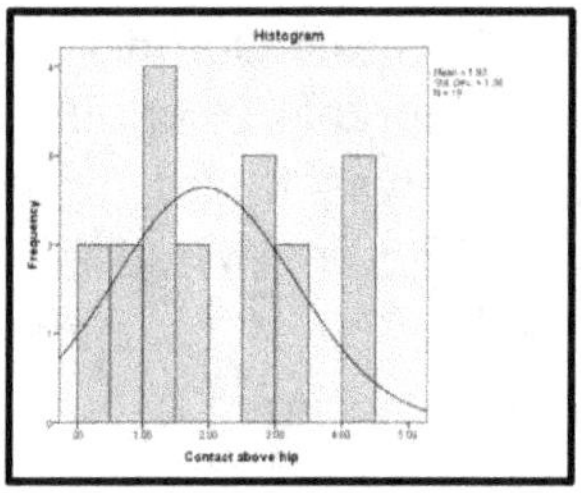

Figure 40 Histogram of contact above hip height in top 10 players

Descriptive statistics were generated for participants with point of contact above hip height in the forehand groundstroke ranked inside the top 10 on either the WTA or ATP tours. The total number of participants with a top 10 ranking is 18. The mean is 1.9 (SD = 1.3) and the range 4. The frequency of point of contact above hip height reported in top 10 participants is at its highest at 1 and 4, occurring 16.7 % of the time; all other numbers are <3, the cumulative percent is 100.0.

When compared against player rankings inside the top 10, the table illustrate that players are making contact above hip height when contacting their forehand groundstroke on average twice out of a possible 30 balls observed. This result indicates that players who hold a top 10 ranking contact above hip height less than 6 % of the time when contacting their forehand groundstroke.

Top 10 V contact at/around hip height

Although height of point of contact at/around hip height aligns with the characteristics of The V by Dr. B ©, it was of interest to identify if the frequency of this variable is as consistent amongst the world's best (top 10) as the overall mean found in Point of Contact. It is also important if this variable is in fact applied less than the overall mean, indicating that the more frequent variable of height of point of contact below hip height, to be the most favoured lower body position among the world's elite group.

When generating frequency statistics, the mean was found to be 6.3 ($SD = 2.4$). This height of point of contact, as illustrated in the histogram, was found to be applied by the elite group (6.3), though when compared to the overall mean of 7.3, the elite group are less likely to contact their forehand groundstroke at this height, see Figure 41.

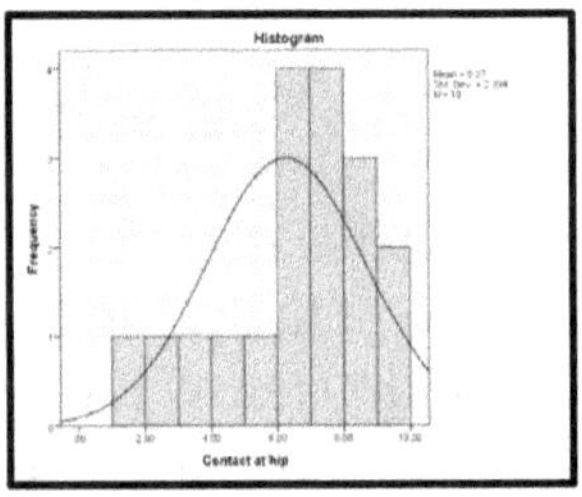

Figure 41 Histogram of contact at/around hip height in top 10 players

Descriptive statistics were generated for participants with point of contact at hip height in the forehand groundstroke ranked inside the top 10 on either the WTA or ATP tours. The total number of participants with a top 10 ranking is 18. The mean reported is 6.3 ($SD = 2.3$) and the range 8.5. The frequency of point of contact at hip height reported in top 10 participants is at its highest at 7, occurring 22.2 % of the time; the frequency of all other results is <4 with a range between >0 but <10, the cumulative percent is 100.0.

When compared against player rankings inside the top 10, the table illustrates that players are making contact around hip height when contacting their forehand groundstroke on average six times out of a possible 30 balls observed. This result indicates that players who hold a top 10 ranking contact around hip height 20 % of the time when contacting their forehand groundstroke.

Top 10 V slight knee bend + contact at/around hip height

When considering the mixed criteria of slight knee bend and height of point of contact at hip height, it was of interest to compare the elite group to the overall mean found to be 2.6. The frequency statistics for the mean was found to be 3.1 (*SD* = 2.2).

Descriptive statistics were generated for participants with slight knee bend collated with point of contact at hip height in the forehand groundstroke ranked inside the top 10 on either the WTA or ATP tours. The total number of participants with a top 10 ranking is 18. The mean is 3.1 (*SD* = 2.2) and the range 8.0. The frequency of slight knee bend collated with point of contact at hip height reported in top 10 participants is at its highest at 2.5, occurring 16.7 % of the time; the frequency of all other results is between 5.6 % and 11.1 %, the cumulative percent is 100.0. The histogram illustrates the tallest point ranging between 2 and 3 representing slight knee bend collated with point of contact at hip height reported in the 18 participants observed at point of contact in the forehand groundstroke.

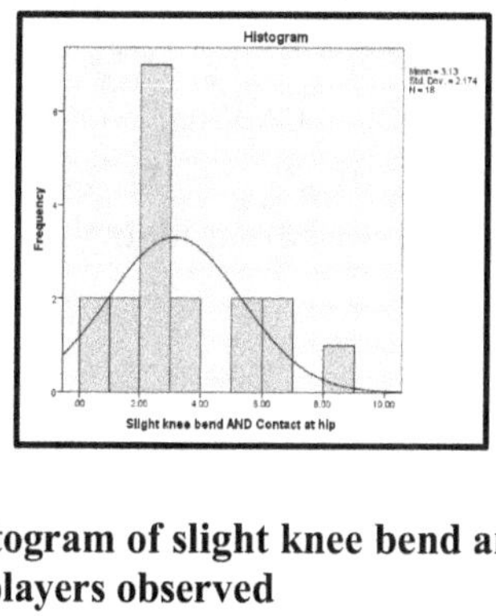

Figure 42 Histogram of slight knee bend and contact at/around hip height in top 10 players observed

When compared against player rankings inside the top 10, the tables illustrate that players are applying slight knee bend whilst contacting around hip height when contacting their forehand groundstroke on average three times out of a possible 30 balls observed. This result indicates that players who hold a top 10 ranking apply this combination 10 % of the time when contacting their forehand groundstroke.

Top 10 V slight bend + contact below hip height

When considering the mixed criteria of slight knee bend and height of point of contact below hip height, it was of concern to compare the elite group to the overall mean, found to be 2.0 (Mixed Criteria: Knee Bend [deep/slight] and Point of Contact [above/hip/below]). The frequency statistics for the mean were 1.7 ($SD = 1.8$).

Descriptive statistics were generated for participants with slight knee bend collated with point of contact below hip height in the forehand groundstroke ranked inside the top 10 on either the WTA or ATP tours. The total number of participants with a top 10 ranking is 18. The mean is 1.6 ($SD = 1.8$) and the range 5.7. The frequency of slight knee bend collated with point of contact below hip height reported in top 10 participants is at its highest at 0.5, 1 and 2 - occurring 11. 1 % of the time (excluding zero); the frequency of all other results is 5.6 % in the range of >0 but <6, the cumulative percent is 100.0. The histogram illustrates the tallest point ranging between 1 and 3 (with zero excluded) representing slight knee bend collated with point of contact below hip height, reported in the 18 participants observed at point of contact in the forehand groundstroke.

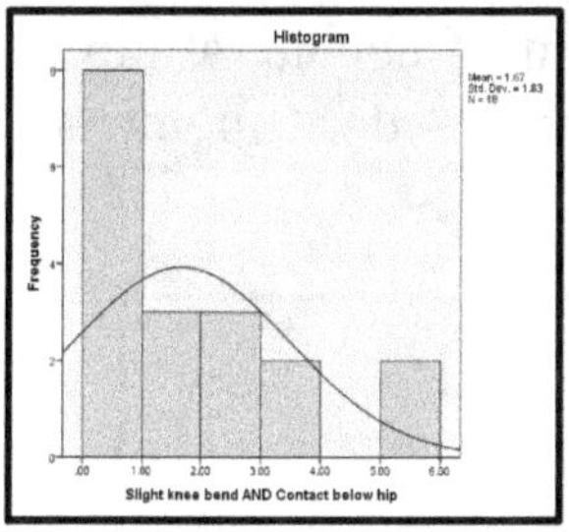

Figure 43 Histogram of slight knee bend and contact below hip height in top 10 players observed

When compared against player rankings inside the top ten, the tables illustrate that players are applying slight knee bend whilst contacting below hip height when contacting their forehand groundstroke on average 1.5 times out of a possible 30 balls observed. This result indicates that players who hold a top ten ranking apply this combination <7 % of the time when contacting their forehand groundstroke.

Top 10 V deep knee bend + contact at/around hip height

When considering the mixed criteria of deep knee bend and height of point of contact at/around hip height, it was again of interest to compare the elite group to the overall mean, found to be 4.7 (Mixed Criteria: Knee Bend [deep/slight] and Point of Contact [above/hip/below]). Through the frequency statistics the mean was found to be 3.3 (*SD* = 1.6).

Descriptive statistics were generated for participants with deep knee bend collated with point of contact at hip height in the forehand groundstroke ranked inside the top 10 on either the WTA or ATP tours. The total number of participants with a top 10 ranking is 18. The mean is 3.3 (*SD* = 1.6) and the range 6.0. The frequency of deep knee bend collated with point of contact at hip height reported in top 10 participants is at its highest at 3.5 and 4, occurring 16.7 % of the time; the frequency of all other results is in the range of >0 but <3, the cumulative percent is 100.0. The histogram illustrates the tallest point ranging between 4 and 5 representing deep knee bend collated with point of contact at hip height reported in the 18 participants observed at point of contact in the forehand groundstroke.

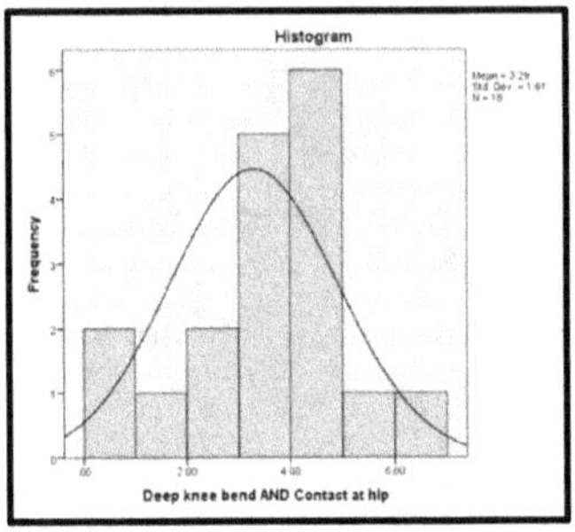

Figure 44 Histogram of deep knee bend and contact at/around hip height in top 10 players observed

When compared against player rankings inside the top ten, the tables illustrate that players are applying deep knee bend whilst contacting around hip height when contacting their forehand groundstroke on average 3 times out of a possible 30 balls observed. This result indicates that players who hold a top ten ranking apply this combination 10 % of the time when contacting their forehand groundstroke.

Gender differences

When generating the descriptive statistics between male and female participants observed, the mean was found as 0.5 (*SD* = 0.5) and the range 1.0. The frequency of male and female participants reported is 60 males and 63 females, 48.8 % and 51.2 % respectively, the cumulative percent is 100.0.

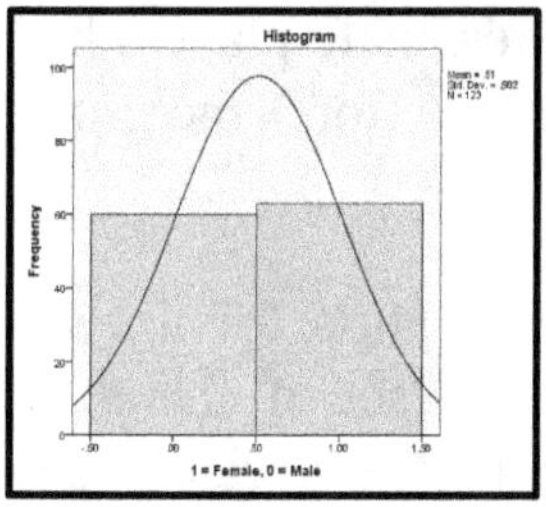

Figure 45 Histogram of gender differences in the 123 players observed

Gender V slight knee bend

Descriptive statistics were generated for gender differences and slight knee bend at point of contact in the forehand groundstroke. The mean is 7.3. Gender differences were correlated against slight knee bend at point of contact in the forehand groundstroke (*SD* = 5.0) and Pearson's correlation found r = 0.84 for slight knee bend with p = 0.357 not of statistical significance (2-tailed). The tables illustrate slight knee bend reported to have an average frequency between >0 but <5 in male participants whilst female participants were reported to have a slightly higher frequency between >8 but <12 at point of contact.

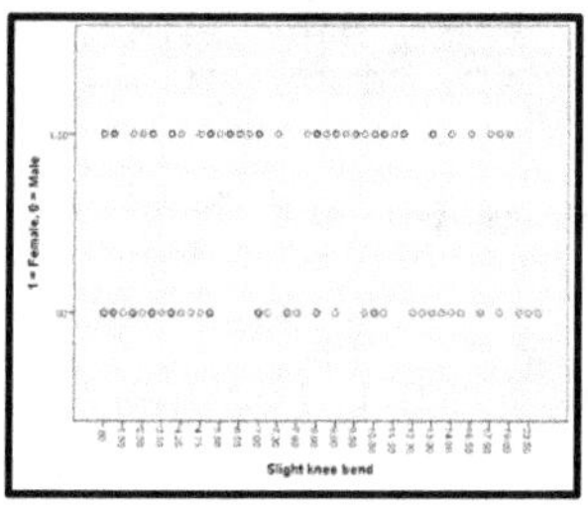

Figure 46 Scatter plot of gender and slight knee bend in the 123 players observed

When compared against gender, slight knee bend is predominantly streamlined on the scatter plot for both male and female players. The scatter plot illustrates that female players have a slightly higher application of slight knee bend than male players; however, male players have a higher range than female players during the forehand groundstroke.

Gender V deep knee bend

Looking at the gender allocation for the variable deep knee bend, the objective was to see if the distribution between males and females due to the variable having the highest frequency of applied knee bend in Knee Bend. Descriptive statistics were generated for gender differences and the mean was found to be 22.7 ($SD = 4.9$). Gender-differences were then correlated against deep knee bend where Pearson's correlation $r = -0.071$ for deep knee bend with $p = 0.437$ not of statistical significance (2-tailed). The scatter plot in Figure 47 illustrates deep knee bend in male participants (slightly higher) whilst female participants were reported to have a slightly lower frequency at point of contact in the forehand groundstroke.

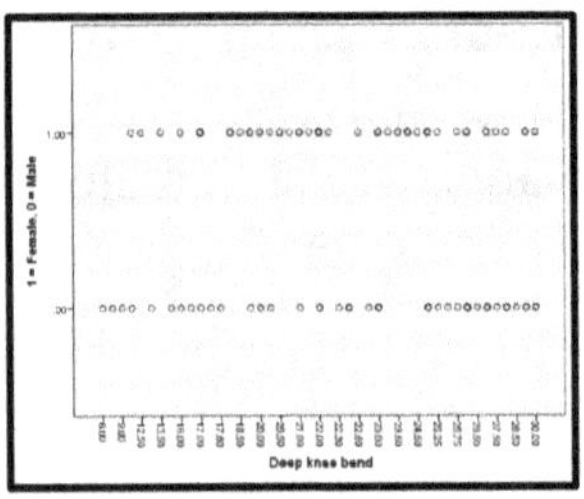

Figure 47 Histogram of gender and deep knee bend in the 123 players observed

The results indicate that when considered singularly, deep knee bend is more frequent amongst male tennis players (>24 but <31), in contrast to female players (>17 but <23). It has been identified that when considering the lesser frequencies, males are also the more frequent, although females had a more consistent range. Looking at the higher range, males are the most dominant when considering >24 balls out of a possible 30, indicating that although more dominant, females appear to have a closer range.

Gender V above contact

Descriptive statistics were generated for gender differences and point of contact above hip height in the forehand groundstroke. The mean is 3.0. Gender differences were correlated against point of contact above hip height in the forehand groundstroke (*SD* = 2.2) and Pearson's correlation found r = 0.237 for point of contact above hip height with p = 0.008 statistical significance (2-tailed) indicating a divide in above between males and females; however, the consistency means illustrate that this level of frequency is not as consistent across genders. The tables illustrate point of contact above hip height reported to have an average frequency between >0 but <2.5 in male participants whilst female participants were reported to have a slightly higher frequency between >2.6 but <6 at point of contact.

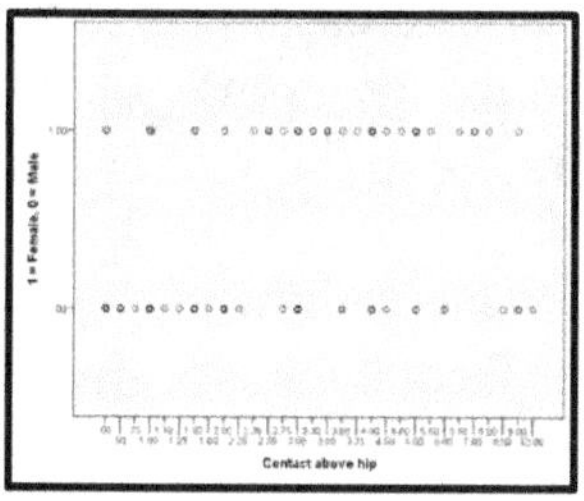

Figure 48 Scatter plot of gender and contact above hip height in the 123 players observed

Gender V hip contact

Descriptive statistics were generated for gender differences and point of contact at hip height in the forehand groundstroke. The mean is 7.3. Gender differences were correlated against point of contact at hip height in the forehand groundstroke (*SD* = 7.3) and Pearson's correlation found r = 0.226 for point of contact at hip height with p = 0.012 statistical significance (2-tailed).

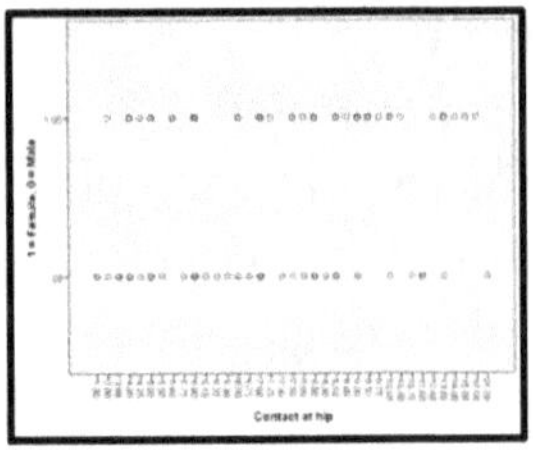

Figure 49 Scatter plot of gender and contact at/around hip height in the 123 players observed

When compared against gender, contact around hip height is predominantly streamlined on the scatter plot for both male and female players. The scatter plot illustrates that females have a higher frequency of contacting around hip height as shown by the grouping towards the end of the range; however, males are shown to have a high frequency of contacting around hip height with an increased range but slightly lower frequency, as shown by the groupings at the beginning and towards midway of the range in the scatter plot.

Gender V contact below hip height

After considering the frequency of males and females contacting the forehand groundstroke below hip height, it was of interest to identify if males had a similar range to that of deep knee bend, and if females continued to slightly differ. Descriptive statistics were generated for gender differences and the mean was found to be 19.6 (SD = 4.0). The correlation of gender-differences identified Pearson's correlation r = -0.304 for point of contact below hip height with p = 0.001 statistical significance (2-tailed). The scatter plot in Figure 50 illustrates point of contact below hip height to have an average frequency between >20.5, but <29 in male participants, whilst female participants had a slightly lower frequency between >10 but <21 at point of contact, indicating a significant result in gender differences.

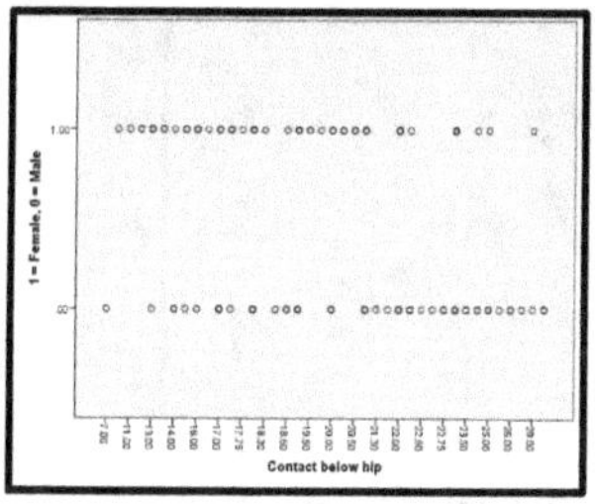

Figure 50 Scatter plot of gender and contact below hip height in the 123 players observed

This result suggests that male tennis players are slightly more consistent when it comes to making contact with the forehand groundstroke below hip height. Females, although found to have a relatively close frequency to their male counterparts, are slightly outweighed by males when considering both variables.

Gender V slight knee + hip contact

Descriptive statistics were generated for gender-differences and slight knee bend collated with point of contact at hip height in the forehand groundstroke. The mean is 2.5. Gender differences were correlated against slight knee bend collated with point of contact at hip height in the forehand groundstroke ($SD = 2.4$), and Pearson's correlation found $r = 0.143$ for slight knee bend collated with point of contact at hip height with $p = 0.114$ not of statistical significance (2-tailed). The tables illustrate slight knee bend collated with point of contact at hip height reported to have an average frequency between >0.5 but <2 in male participants whilst female participants were reported to have a slightly higher frequency between >2 but <4 at point of contact.

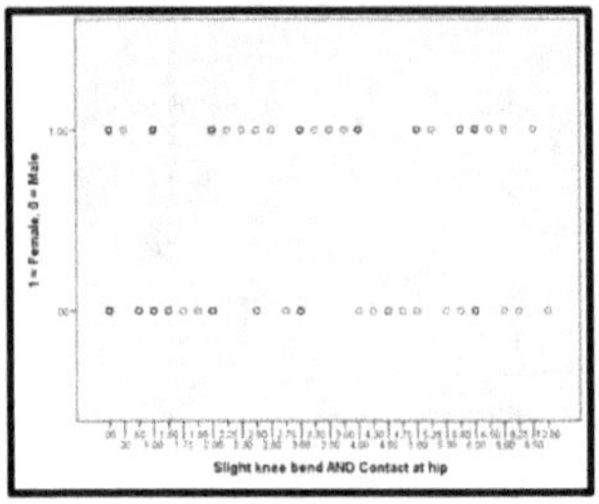

Figure 51 Scatter plot of gender and contact at/around hip height in the 123 players observed

When compared against gender, contact around hip height and slight knee bend is predominantly streamlined on the scatter plot for both male and female players. The scatter plot illustrates that males have a slightly higher frequency of applying the combination as shown by the grouping towards the end of the range; however, females are shown to have a slightly lower frequency of applying the combination as shown by the groupings midway on the scatter plot.

Gender V slight knee + below contact

Descriptive statistics were generated for gender differences and slight knee bend collated with point of contact below hip height in the forehand groundstroke. The mean is 2.0. Gender differences were correlated against slight knee bend collated with point of contact below hip height in the forehand groundstroke (*SD* = 3.3), and Pearson's correlation found r = -0.174 for slight knee bend collated with point of contact below hip height with p = 0.055 not of statistical significance (2-tailed). The tables illustrate slight knee bend collated with point of contact below hip height reported to have an average frequency between >3 but <5 in male participants, whilst female participants were reported to have a slightly lower frequency between >1 but <2.5 at point of contact.

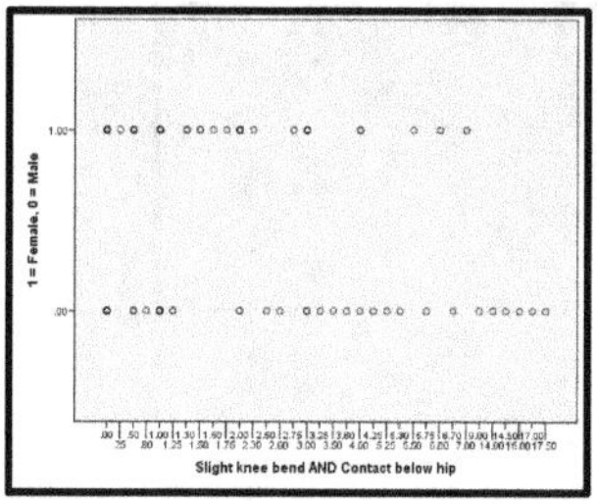

Figure 52 Scatter plot of gender and slight knee bend AND contact below hip height in the 123 players observed

When compared against gender, contact below hip height and slight knee bend is more frequent in males than females on the scatter plot. The scatter plot illustrates that males have a higher frequency of applying the combination, as shown by the grouping towards the end of the range, whilst females are shown to have a lower frequency of applying the combination as shown by the groupings predominantly at the beginning on the scatter plot.

Gender V deep knee + hip contact

Descriptive statistics were generated for gender differences and deep knee bend collated with point of contact at hip height in the forehand groundstroke. The mean is 4.7. Gender differences were correlated against deep knee bend collated with point of contact at hip height in the forehand groundstroke ($SD = 3.1$), and Pearson's correlation found $r = 0.105$ for deep knee bend collated with point of contact at hip height with $p = 0.248$ not of statistical significance (2-tailed). The tables illustrate deep knee bend collated with point of contact at hip height reported to have an average frequency between >1 but <4 in male participants whilst female participants were reported to have a slightly higher frequency and wider range between >2 but <14 at point of contact.

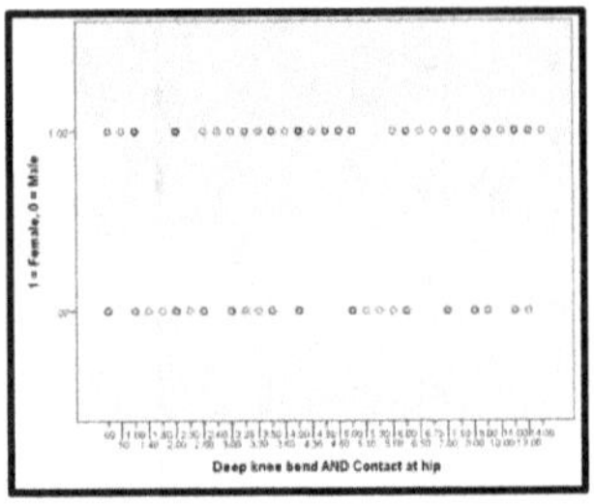

Figure 53 Scatter plot of gender and deep knee bend AND contact at/around hip height in the 123 players observed

When compared against gender, contact around hip height and deep knee bend is predominantly streamlined on the scatter plot for both male and female players. The scatter plot illustrates that females have a slightly higher frequency of applying the combination, as shown by the grouping towards the end of the range, whilst males are shown to have a slightly lower frequency of applying the combination as shown by the groupings midway on the scatter plot and towards the beginning. No clear pattern is identified.

Observation method

When generating the frequency statistics, each method was labelled 1, 2 or 3 to illustrate the shared notational analysis and if one was favoured over the other. The number 1 represents viewed live, 2 equals viewed footage, and 3 is used for both options in the participants observed during the contact phase of the forehand groundstroke. Overall, the frequency of viewed live or viewed footage was found to be 50.0 and 50.0 respectively, a clear even distribution between the two methods of notational analysis. The frequency of 3 was 23, 18.7% of the cumulative percent, indicative of both options having an even distribution. The histogram in Figure 54 illustrates the frequency in viewed live or viewed footage with a 40.7% share each of the 123 participants observed at point of contact in the forehand groundstroke.

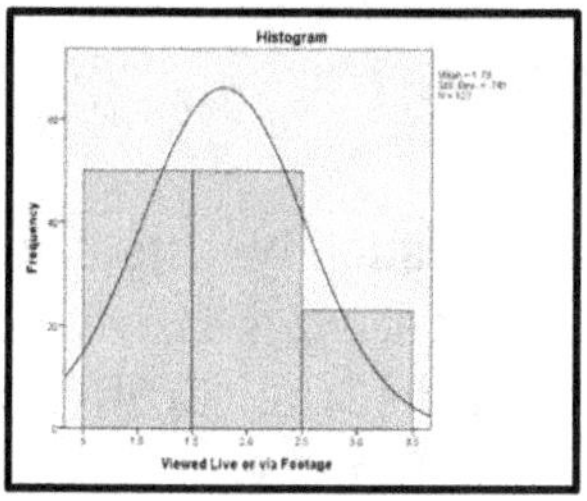

Figure 54 Histogram of viewed live or via footage in the 123 players observed

Although the observation method was not controlled, the intention was to ensure an objective point of view was retained throughout the notational analysis and that neither method outweighed the other. Interestingly, due to multiple players having their overall scores averaged, a result of being observed several times (a consequence of winning performances), it is significant that the overall distribution of results has maintained an even spread over the three methods of observation.

Part 3: Conclusion

The *less* than statistically significant criteria – individually and mixed, *adds* clarity and conviction to *The Unveiling* and its statistically significant data, disclosed for the very *first* time. Whilst *The Statistics* were rudimentary in their initial release, their profiling is indicative of one of many layers unfolded to reveal the *truth* inside, and behind elite performance parameters.

On face value the numerals are often undecipherable to the non-*scientist* or with null *real-world* implications. *The Statistics* were shared to the contrary – to provide further scaffolding of the *what* variations and their limited statistical inferences, further compounding the disclosed data in *The Unveiling.*

Scientific developments are arduous, enduring and often *less* than statistically significant. Yet, it is the rigour of the scientific *process* that establishes the new literature and its consequential scientific discovery – propelling the world as we know it. After all, *science* makes the world go 'round.

Whilst statistical inferences are *one* half, *real-world* application are the other. Without applied cause-effect relationships, the *discovery* would remain idle. Recall, the autonomy loop. *The Statistics* further cement the implications of The *Unveiling*, whilst *Coaching Analyses* is cause-effect in action – The V by Dr. B © is very much *alive* and *applied* in reality, courtesy of the *one-percenters* and their integration of its transference in their autonomy loop.

Data is often overwhelming as a result of its scale. *The Statistics* is a mere end-point that has digressed the implications of the 121,770 individual notations – in direct association with *The Unveiling*, pulled a part in its entirety, to relay scientific discovery. The criteria inside *The Statistics* shared *less* than optimal data – complementing the now disclosed statistically significant data that *unveiled* The V by Dr. B ©.

CONCLUSION

The inevitability of change is synonymous with time. Whether yearned for or the want to preserve, time continues – *change* occurs. Tennis is *no* different. The game has now *forever* changed. Courtesy of scientific discovery, tennis has reached new heights and opened up a whole new ball game – the *top 10* for the taking. Climbing to the pinnacle of tennis and becoming a *one-percenter* has never been more streamlined in the history of sport – The V by Dr. B © has solidified its historical implications for athletes the world round. But, failure is mutually on the cards.

Functional movement parameters are key performance indicators. Likewise, varied technical parameters with their respective biomechanical arrangements have been adapted over time for a technique to prevail – fluid or not. The V by Dr. B © gives the power back to the player and coach, respectively, to put a stop to the ailing dominance of injury across the sporting spectrum – functionality is key. Synonymous with functional movement parameters is the dual introduction of The V by Dr. B © and The Rule of Transference – together the truth behind a *one-percenter* is translucent. In conjunction with *The Optimal Performance Theory* (Berge, 2017) never before has the attainment of a *one-percenter* been laid out in print *nor* been the end-point of scientific discovery. The landmark discovery of The V by Dr. B © and its technical implications for tennis players is profound – yet its expansion courtesy of The Rule of Transference has just begun.

The playing generation of the past *seven* years (2012—2019) is by large responsible for the significant data compiled – 121,770 individual notations, and the *ninety-nine* individual scientific breakdowns in alignment with The V by Dr. B © and The Rule of Transference. Publicly accessible at all times, individual player accessibility varied – yet the overall cohort (1—200) of *one-hundred* males and *one-hundred* females at any given time for the *Coaching Analyses*, broadened to *two-hundred* for *The Statistics* strengthened player availability for a denser data set. The range of *one-percenters* was always going to be scarce – with *ten* males and *ten* females inclusive at any given time; *less* with the restrictions imposed by accessibility.

The scientific implications disclosed in *The Unveiling* are ground breaking of the near *seven* years endured – disclosing in print the **statistically significant** data sandwiched with its derived significance, *The Statistics*. Yet, real-world application signifies transference of scientific discovery into reality *and* its practicality in cementing optimal performance parameters – the criteria. Perpetuated in the *Coaching Analyses*, The V by Dr. B © and its *consistent* application was fundamental, synonymous with a *one-percenters* performance – yet conditional of The Rule of Transference.

Recall, The Rule of Transference works hand-in-hand with The V by Dr. B © whereby the functional values of The V by Dr. B © and the resultant technical parameters are key – limitations in applying The Rule of Transference delimits a player, statistically, of securing and maintaining a *one-percenter* advantage. The presence and consistent application of angular movement is thus afforded to players who apply The Rule of Transference. Likewise, angular movement is more readily applied by players prior to The V by Dr. B © – the angular stance further excels its properties (the criteria); yet, angular movement *without* The V by Dr. B © loses its stronghold. Angular movement *with* The V by Dr. B © has now been disclosed as an optimal performance parameter of the *one-percenters* – applying The Rule of Transference.

Acceptance comes *first* – akin to the *truth* behind scientific discovery. The mathematical equations of statistically significant data are truth by numbers. However, human psychology denotes the struggle of acceptance of the new *if* it impinges on the current set of values. Recall, *change is inevitable – synonymous with time – tennis was the sport of choice dismantled*; it can't be undone. Scientific discovery affords the new – providing the answer to the elusive question: **What is Your Game Missing?** The V by Dr. B © is unrivalled science – *without* coincidence, *with* a mathematical undertone and literal *hundreds* of *thousands* of notations. Yet, human psychology is finicky as aforementioned – an answer in itself that either catapults or hinders those from, or into the *one-percenter* niche.

A generation is now forever changed. The V by Dr. B © coupled with The Rule of Transference delimits the onset of injury – no more does performance come with the consequence of a flailing body. Power and speed can be relished and brought to new heights through The V by Dr. B © without compromise of functional movement – *your* body is now on *your* side. Longevity is possible – no longer will players be at the mercy of techniques that do not account for their body as a whole in correspondence with optimal performance parameters. Scientifically, it now exists – it is now possible to become a *one-percenter*, knowing the answer behind **What is Your Game Missing?** Literally, becoming a *one-percenter* has never in the history of sport been so accessible.

Historically, a new era has dawned on the tennis world. Elitedom is now closer than ever before. The *answer* to **What is Your Game Missing** is now transparent and The V by Dr. B © is the beginning of the next chapter to beckon the tennis – and sporting world. The Rule of Transference and the historical scientific discovery of The V by Dr. B © – *missing* until now, is *the* new dawn – the signal of a *new* level of play, unbeknownst to now – a broken twig holds the key.

WHAT NOW?

Visit https://AM8International.com and https://Topicthread.com for what's happening now and the latest resources available.

Want to get in touch? You can find Dr. B on Topicthread @Dr.B, or on limited social platforms @dr_ashley_b where she shares the latest news from Topicthread as well as her current running endeavours, publications, and performances.

BOOKS BY DR. B

The Secrets to Optimal Performance Success
(March, 2016)

The Secrets to Optimal Wellbeing
(September, 2016)

The Science of Elite Performance
(March, 2017)

The Secrets to Optimal Coaching Success
(March, 2018)

The Elite Research Method
(October, 2018)

What is Your Game Missing?
(March, 2019)

What is Your Game Missing, Now?
(October, 2019)

What is Your Game Missing, to Win?
(July, 2020)

I am Your Tennis Coaching Guru
(December, 2020)

The 7 Keys to Optimise Your Life
(December, 2021)

BIBLIOGRAPHY

Berge, A.M (2014). How the V Position is Applied and Communicated in the Forehand Groundstroke by Elite Players and Coaches Internationally. Griffith University: Gold Coast, QLD.

Berge, A.M (2016). The Secrets to Optimal Wellbeing: a 12 week freestyle nutrition and functional movement guide. AM8 International: Gold Coast, QLD.

Berge, A.M (2017). The Science of Elite Performance: the world awaits. AM8 International: Gold Coast, QLD.

Garhammer, J. (1989). Biomechanics I. In P. J. Rasch (Ed.), *Kinesiology And Applied Anatomy* (7th ed.) (pp. 78–101). Philadelphia, PA: Lea & Febiger.

Gratton C., & Jones, I. (2010). *Research Methods for Sports Studies*. London UK: Routledge.

Hughes, M., & Franks, I. (2004). *Notational Analysis of Sport: Systems for Better Coaching and Performance in Sport* (2nd ed). New York, NY: Routledge.

Long, J. (2007). *Researching Leisure, Sport and Tourism – The Essential Guide*. London, UK: Sage Publications.

Network 10 Australia (2013). Hopman Cup 2013 [Television Broadcast] In. *Ten Sport.* Pyramont, Australia: Network 10 Australia

Schonborn, R. (2000). What is technique? *Advanced Techniques for Competitive Tennis* (2nd ed.) (pp. 11–72). Oxford, UK: Meyer & Meyer Sport.

Seven Network Australia (2013). The Australian Open 2013 [Television Broadcast] In. *Seven Sport.* Pyramont, Australia: Seven Network Australia

Viera, A.J., & Garrett, J.M. (2005). Understanding interobserver agreement: the kappa statistic. *Family Medicine, 37*(5), 360-363.

www.ingramcontent.com/pod-product-compliance
Lightning Source LLC
La Vergne TN
LVHW052300200926
848481LV00002B/417
9780994569479